Volume 26

Ceramic Transactions

FORMING SCIENCE AND TECHNOLOGY FOR CERAMICS

Edited by Michael J. Cima, Massachusetts
Institute of Technology

The American Ceramic Society
Westerville, Ohio

Proceedings of the Symposium on Forming Science and Related Properties of Ceramics held during the 93rd Annual Meeting of the American Ceramic Society in Cincinnati, OH, April 29–May 3, 1991.

Library of Congress Cataloging-in-Publication Data

American Ceramic Society. Meeting (93rd : 1991 : Cincinnati, Ohio)
    Forming science and technology for ceramics / edited by Michael J. Cima.
        p.   cm. -- (Ceramic transactions ; v. 26)
    "Proceedings of a symposium ... held during the 93rd Annual Meeting of the American Ceramic Society in Cincinnati, OH, April 29-May 3, 1991"--T.p. verso.
    Includes bibliographical references and index.
    ISBN 0-944904-48-3
    1. Ceramics--Congresses.   I. Cima, Michael J. II. American Ceramic Society.   III. Title. IV. Series.
TP786.A47   1991
666--dc20   91-40343
                                                                       CIP

Printed in the United States of America
1   2   3   4—96   95   94   93   92

ISBN 0-944904-48-3

Preface

The commercialization of advanced ceramic components has been hindered by expensive and lengthy production processes and, increasingly, ceramics forming technology is the crucial limiting factor in the development of these advanced ceramics. Similar barriers exist for other advanced materials. A recent report by the National Research Council, "Materials Science and Engineering for the 1990s: Maintaining Competitiveness in the Age of Materials,"[1] concludes that processing technology is emerging as a coherent field in many materials areas, but that the United States shows a serious weakness in applying basic science to the processing and manufacturing of advanced materials. The Symposium entitled "Forming Science and Technology for Ceramics" at the 1991 Annual Meeting of the American Ceramic Society was organized to discuss specific examples of how basic processing science is being applied to ceramics forming. It is hoped that the publication of this proceedings volume will stimulate more basic research on forming technology.

Ceramic forming methods are now considered to be the "enabling" technology for producing advanced structural and electronic materials. Commercial acceptance of advanced ceramic components hinges on the ability to produce defect-free complex shapes with high dimensional tolerance. A primary objective of this proceedings was to solicit meaningful contributions from academia and industry which would provide an assessment of advanced forming science practice. Contributions addressed both electronic and structural applications of ceramics, since these areas share many of the same issues, such as state of particle dispersion, choice of chemical processing aids, and the effects of deformation stress in forming the green ceramic. In addition, articles focused on unit processes related to forming, such as pressure casting, injection molding, drying, processing aid removal, and green machining. Contributions mark the industry's progress and the development of our understanding since the publication of the last proceedings on ceramics forming technology in 1983. Finally, a significant number of articles deal with applications of basic science to forming issues, such as colloid science, rheology, chemistry, transport, and mechanics. Thus, this proceedings summarizes basic and applied processing research during the 1980s and will perhaps help policymakers in government and industry set goals for the next decade.

New concepts in ceramic component manufacturing are emerging as markets develop for advanced ceramics. It is clear the primary motivation for development of colloidal processes for ceramics forming was the desire to eliminate microstructural flaws.[2] Acceptance of these technologies, however, may depend on their ability to reduce development time and provide for flexibility in manufacturing. An important example is low pressure molding, on which there are several articles in this proceedings.[3,4,5] These techniques permit the

---

(1) National Research Council, Materials Science and Engineering for the 1990s: Maintaining Competitiveness in the Age of Materials. National Academy Press, Washington D.C., 1989.

(2) H.K. Bowen, "Basic Research Needs on High Temperature Ceramics for Energy Applications," *Mater. Sci. Eng.*, **44** 1-56 (1980).

(3) O.O. Omatete, R.A. Strehlow, and C.A. Walls, "Drying of Gelcast Ceramics"; this book, pp. 101-07.

use of soft tooling, which can greatly reduce the time and costs involved in the production of a ceramic component. Manufacturing strategies based on flexible tooling approaches are important for advanced ceramics, where production volume is usually small and where custom production can add significant value to the component.

The ideal flexible manufacturing process for ceramics would require only such information as the dimensions, surface finish, and microstructure of the component while imposing no process constraints on the component design. Unfortunately, no such "smart" processes currently exist in practice. Injection molding, for example, requires that the thickness of component sections be roughly constant throughout the component to ensure uniform shinkage upon firing.[6] In addition, the use of a mold requires that allowances be made for the demolding process, and this imposes topological constraints on component design.

In the future, forming technology will no doubt capitalize on the modern computer-aided-design (CAD) environment to fabricate real objects of complex shape. Software has already been developed to provide a link between solid models and NC machine centers, for example. More recently, processes that build complex shapes in layers are becoming available. Stereolithography,[7] for example, constructs successive layers by writing on the surface of a bath containing a UV-curable resin. The trajectory the laser takes on the surface is determined by a computer, which mathematically slices a model of the object. Sequential layers are built on top of one another to construct the component. Alternatively, powder processes such as selective laser sintering (SLS)[8] and three dimensional printing (3DP)[9] function in a similar way but may be more applicable to ceramics. The latter, in fact, has been used to build complex refractory structures for metal castings.[10] In three dimensional printing, layers of powdered material are deposited and selectively bound by a binder deposited via ink-jet printing. After all the layers are complete, the unbound powder is removed and the bound powder forms the walls of a three dimensional part. This process is well suited to ceramics and some metal alloys that are normally processed via a particulate state.

As with many scientific meetings, important ideas surfaced in discussions after presentations and in the hallways of the conference center. Although it is impossible to summarize all the interesting dialogue which took place during the course of the meeting, I am hopeful

(4) B.E. Novich, C.A. Sundback, and R.W. Adams, "Quickset™ Injection Molding of High Performance Ceramics"; this book, pp. 157-66.

(5) M.J. Edirisinghe, K.L. Tomkins, and M. Patching, "Alcohol Based Binder Systems for Moulding Ceramic Materials"; this book, pp. 165-71.

(6) J.R. Merhar, "Designing for Metal Injection Molding"; presented at the Powder Injection Molding International Symposium, July 15-17, 1991, Albany, NY; see also Inject Alloy® product literature, DuPont, Wilmington, DE.

(7) T. Wohlers, "Creating Parts by Layers," *Cadence*, 5 73-76 1989.

(8) C. Deckard and J. Beaman, "Recent Advances in Selective Laser Sintering," pp. 447-52 in Proceedings of the 14th Conference on Product Research and Technology, University of Michigan, October 1987.

(9) E. Sachs, M. Cima, P. Williams, and D. Brancazio, "Rapid Tooling and Prototyping by Three Dimensional Printing," *Trans. NAMRI/SME*, 1990, 41-45.

(10) M.J. Cima and E.M. Sachs, "Three Dimensional Printing: Form, Materials, and Performance"; in Proceedings of the Solid Freeform Fabrication Symposium, Austin, TX, 1991, in press.

this proceedings documents most of the significant analyses and innovations presented at the Symposium.

I wish to thank the American Ceramic Society for their decision to hold a symposium on forming. I am also grateful for the advice I received from Prof. Michael Sacks at the University of Florida. The financial support of MIT's Ceramics Processing Research Laboratory (CPRL) is gratefully acknowledged. The publication of this proceedings would not have been possible without the tireless efforts of Ms. Eve Downing, the CPRL technical editor.

Michael J. Cima
Cambridge, Massachusetts
September 1991

Ceramic Transactions is a new proceedings series designed to meet two needs: high quality content and rapid publication. Volumes in the series come from meetings, symposia, and forums. Each paper is reviewed by two peers, and final manuscripts are prepared by authors in a "camera-ready" format. The volumes in this series would not be possible without the hard work, dedication, and cooperation of editors, reviewers, and authors, who all deserve a great deal of thanks.

Your comments, questions, and suggestions for future *Ceramic Transactions* volumes are welcomed and should be addressed to the Director of Publications, The American Ceramic Society, Inc.

Contents

Precipitation of Uniform Particles: General Lessons Derived from Titanium Alkoxides 1
J-L. Look and C.F. Zukoski

Precipitation of Oxalates from Homogeneous Solution: Synthesis of BaTi(C$_2$O$_4$)$_2$ and Ba, Y, and Cu Oxalates 8
W.E. Rhine, R.B. Hallock, W.M. Davis, and M.J. Cima

Alkoxide Synthesis of Al$_2$O$_3$ and Y$_3$Al$_5$O$_{12}$ Powders 17
J. McKittrick, K. Kinsman, S. Connell, E. Sluzky, and K. Hesse

The Effect of Yttria Stabilizer on the Electrokinetic Behaviour of ZrO$_2$-Al$_2$O$_3$ Colloidal Suspensions 24
D. Goski, J.C.T. Kwak, and K.J. Konsztowicz

A Surface Chemical Technique for Sintering Aid Addition 31
S.G. Malghan, A. Sivakumar, and P.S. Wang

Analysis of Surface Chemistry of Silicon Nitride and Carbide Powders 38
S.G. Malghan, P.T. Pei, and P.S. Wang

The Role of Surface Tension in the Formation of Donut-Shaped Granules during Spray-Drying 46
K.J. Konsztowicz, G. Maksym, T. Maksym, H.W. King, W.F. Caley, and E. Vargha-Butler

In-Situ Light Scattering Study of Aggregation 54
Y.H. Rim, J.D. Cawley, R.R. Ansari, and W.V. Meyer

Rheological Detection of Agglomerates in Concentrated Alumina Slurries 66
B.M. Moudgil and M.E. Springgate

The Effect of Dispersant Concentration on the Cast Cake Structure and Rheology 73
I. Tsao and R.A. Haber

Dilatant Transitions of Alpha Alumina Slips 81
D.A. Barclay

Drying Stresses in Granular Ceramic Films 88
R.C. Chiu and M.J. Cima

**Effective Percolation Limit in the Drying of Ceramic
Composites** 95
M.W. Weiser and K.C. Key

Drying of Gelcast Ceramics 101
O.O. Omatete, R.A. Strehlow, and C.A. Walls

**Why Cracks Appear When the Constant Rate Period
Stops in a Drying Process** 108
X. Li

**Deformation during Binder Removal from Multilayer
Ceramic Greenware** 115
Y. Tang and M.J. Cima

**Solid Oxide Fuel Cell Ceramics through Colloidal
Processing** 125
K. Kendall

Numerical Simulation of the Extrusion of Plastic Bodies . . 132
W.B. Carlson, J. Zheng, and J.S. Reed

**Superabsorbent Polymers as Templates in Forming
Oxide Ceramics** 141
A.B. Hardy and W.E. Rhine

Processing of Foamed Ceramics 149
W.P. Minnear

**Quickset™ Injection Molding of High Performance
Ceramics** 157
B.E. Novich, C.A. Sundback, and R.W. Adams

**Alcohol Based Binder Systems for Moulding Ceramic
Materials** 165
M.J. Edirisinghe, K.L. Tomkins, and M. Patching

**Processing of a High Toughness Silicon Nitride
Material** 172
B.J. Meenan, R.A. Haber, and D.E. Niesz

Sintering of Alumina and Zirconia Greens Obtained via
Slip Casting and Pressure Slip Casting 178
A. Salomoni, I. Stamenkovic, A. Tucci, and L. Esposito

Kinetics and Mechanics of Constant Pressure Filtration
of Colloidal Ceramic Dispersions 187
B.V. Velamakanni and F.F. Lange

New Processing Method of Near Net-Shaped,
Complex-Shaped Structural Ceramics 189
B.V. Velamakanni and F.F. Lange

Cellulose Ethers in Tape Casting Formulation 191
K.E. Burnfield and B.C. Peterson

The Effects of Solvents and Binders on the Properties of
Tape Casting Slurries and Green Tapes 197
J-C. Lin, T-S. Yeh, C-L. Cherng, and C-M. Wang

Correlation of Conformation of Acrylic Polymers in
Aqueous Suspensions and Properties of Alumina
Green Sheets 205
K. Nagata

High Performance Electronic Substrates from Tape
Casting of Beryllium Oxide 211
J.L. Sepulveda and R.E. Kottman

High-Speed Precise Electron Beam Perforation
Technology 219
K. Sakurai, Y. Yamane, H. Murakami, S. Sasaki, and S. Takeno

A Method for Making Grooves with Sharp Corners on a
Green Ceramic Body Using a Tool with Biaxial
Ultrasonic Vibration 225
K. Suzuki, T. Uematsu, H. Nakabayashi, and S. Mishiro

Optimizing Green Machining after Isopressing of
Beryllia Ceramic Bodies 231
G.A. Kovell and J.L. Sepulveda

Reliable Electrophoretic Mobility Measurement for
Ceramic Powders 240
J-F. Wang, R.E. Riman, and D.J. Shanefield

PRECIPITATION OF UNIFORM PARTICLES: GENERAL LESSONS DERIVED FROM TITANIUM ALKOXIDES

J-L. Look and C. F. Zukoski
Department of Chemical Engineering and Beckman Institute,
University of Illinois, Urbana, Illinois 61801

INTRODUCTION

Despite the availability of numerous routes to preparing uniform particles from a wide range of chemistries (the work of Matijevic and co-workers should be consulted for a wonderful array of remarkably sized and shaped micron sized inorganic particles[1-3]), little is understood about the underlying physical chemistry that gives rise to precipitate uniformity. Conventional wisdom holds that precipitate size and shape are critically dependent on the details of the reaction chemistry. As a result, the conditions giving rise to, for example, silica spheres derived from silicon alkoxides will have little bearing on the conditions giving rise to rods of FeOOH derived from $FeCl_2$. In this paper, we discuss recent studies indicating that while the internal bonding chemistry will be governed by chemical details, precipitation of uniform particles results from physico-chemical considerations which can be manipulated to advantage.

We use as an example system the precipitation of hydrous titania particles from tetraethylortho-titanate (TEOT) in aqueous ethanol solutions. Uniform particles have been reported for this system by Barringer and Bowen[4,5] but the synthesis routes were found difficult to reproduce by Jean and Ring,[6,7] Edelson and Glaeser,[8] and Harris and Byer.[9] Jean and Ring[6,7] demonstrated that the use of hydroxypropyl cellulose as a steric stabilizer resulted in uniform particles with less sensitivity to reaction conditions. Edelson and Glaeser[8] argue that following the route of Jean and Ring results in particles composed of agglomerates of 2-6 nm diameter subunits. More recently, Bailey and Mecartney,[10] using a cryo-TEM technique, arrested particle growth and observed that even in the absence of hydroxypropyl cellulose, aggregation is a major growth pathway.

Initially, in our studies of the precipitation of titania particles, we were unable to reproduce the results of Barringer and Bowen.[4,5] Only when the importance of

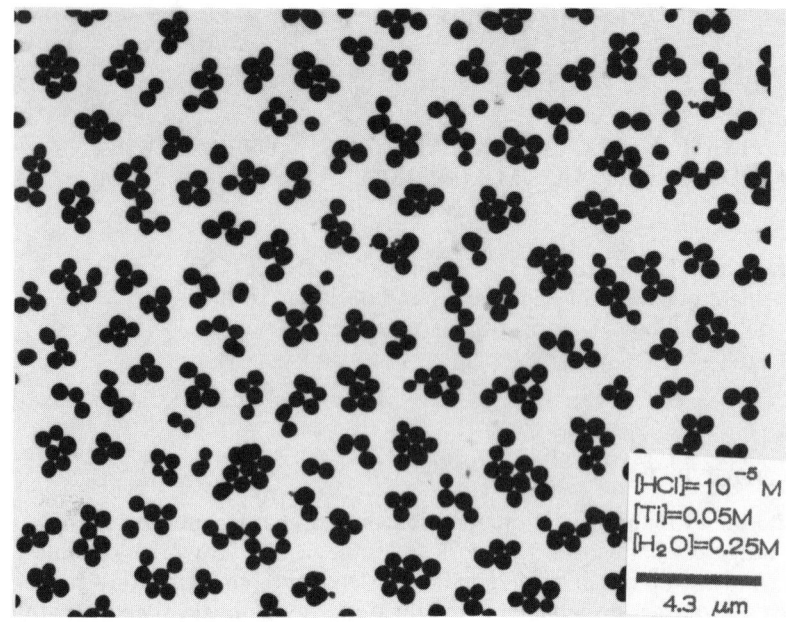

[HCl]=10^{-5} M
[Ti]=0.05M
[H₂O]=0.25M

4.3 μm

Figure 1a Particles precipitated from 0.05 M TEOT, 0.25 M H$_2$O and 10^{-5} M HCl at 25°C in the absence of shear.

colloidal stability during the precipitation reaction was understood were we able to prepare uniform precipitates under conditions reported by these authors. Typical precipitate morphologies, prepared in the absence of steric stabilizers are composed of agglomerates of relatively uniform particles that are fused together by well formed necks (fig. 1b). Only when the level of agitation in the reactor is controlled and the colloidal interaction potential is sufficiently repulsive can uniform precipitates be formed (fig. 1a).[11-13] All uniform precipitates prepared had surface potentials greater than 12 mV in their mother liquors (0.05 M ≤ [TEOT] ≤ 0.2 M, 0.25 M ≤ [H$_2$O] ≤ 1 M, 0 M ≤ [HCl] ≤ 10^{-3} M, 0 ≤ [NaCl] ≤ 10^{-2} M). As the absolute water concentration was raised, the surface potential dropped and agglomerated morphologies were observed. Upon addition of HCl or NaCl surface potentials were raised and higher water concentrations could be used to precipitate uniform particles. The origin of the morphological differences shown in fig. 1 is the focus of this paper.

COLLOIDAL STABILITY DURING PRECIPITATION REACTIONS

Typical reaction conditions involving TEOT result in particles in the 100-1200 nm

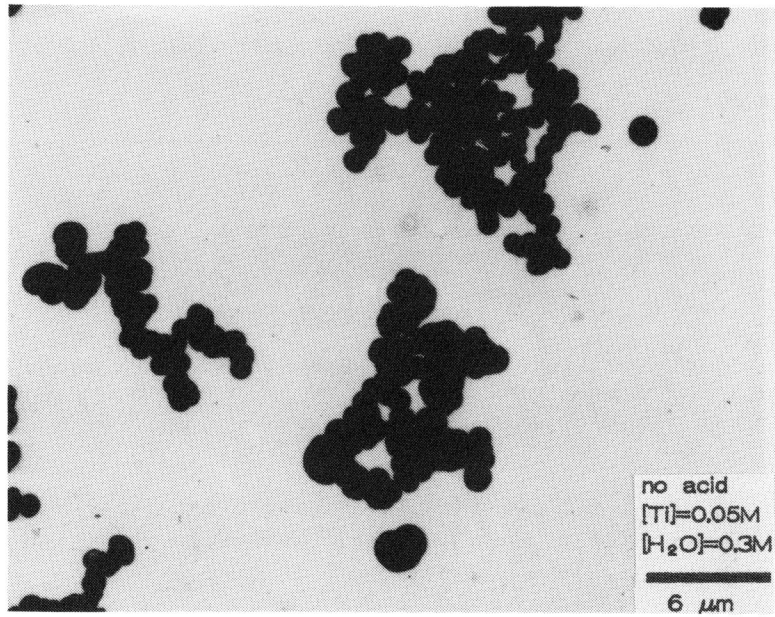

no acid
[Ti]=0.05M
[H₂O]=0.3M

6 μm

Figure 1b Particles precipitated from 0.05 M TEOT and 0.3 M H$_2$O at 25°C in the absence of shear.

diameter size range. The agglomerated morphologies shown in fig. 1 are often composed of particles that appear to have grown as discrete subunits to diameters on the order of several hundred nanometers prior to agglomerating and fusing. This is particularly noticeable in shear experiments where the precipitating particles have surface potentials greater than 12 mV. Here discrete particles grow to a distinct size prior to being agglomerated by the shear. The question we address here is why do colloidally stable particles grow to a particular size where agglomeration occurs?

Conventional colloid stability theory indicates that particles interacting with van der Waals attractive and electrostatic repulsive forces become more stable to aggregation as the surface potential is raised or the ionic strength is decreased. A convenient measure of stability comes through the stability ratio, W_{ii} which is the ratio of the rate of rapid aggregation (determined for particles which feel no attractive or repulsive interactions) to the rate of aggregation in the presence of interactions. W_{ii} can be approximated from[14]

$$W_{ii} = 2\int_{2a}^{\infty} \frac{\exp(V_T/kT)}{r^2}\, dr \qquad (1)$$

where V_T is the total pair interaction energy and r is the center to center pair separation. Classically considered interaction potentials for two spheres of radius a include van der Waals attractive, V_a, and electrostatic repulsive, V_e terms which have forms[14]

$$V_a = \frac{-A}{6}\left[\frac{2}{R^{*2} - 4} + \frac{2}{R^{*2}} + \ln\left(\frac{R^{*2} - 4}{R^{*2}}\right)\right]$$ (2)

and

$$V_e = 2\pi\varepsilon\varepsilon_0\, a\, \psi_0^2 \ln\{1 + \exp[a\kappa\,(R^* - 2)]\}$$ (3)

where A is the Hamaker constant, $\varepsilon\varepsilon_0$ is the product of the dielectric constant of the continuous phase and the permittivity of free space, ψ_0 is the surface potential, κ is the Debye-Huckel parameter determined by ionic strength, and $R^* = r/a$ where r is the pair center to center separation.

If one starts with a particle density of ρ_0 (for the systems studied here $\rho_0 = 10^{16}$-10^{17} m^{-3}), the time to halve the number of particles by Brownian aggregation can be found from[14]

$$t_{1/2} = \frac{3\eta\,W_n}{4kT\rho_0}$$ (4)

where η is the continuous phase viscosity, and kT is the product of Boltzmann's constant and the absolute temperature. Using $A = 2.5\,kT$ (T=25°C), $\kappa = 1.5 \times 10^7$ m^{-1} (as determined from the conductivity of the mother liquor) and $\psi_0 = 12$ mV for 1 μm particles precipitated from 0.05 M TEOT and 0.25 M H_2O, if V_T is calculated as the sum of electrostatic and van der Waals interaction forces, $t_{1/2}$ is on the order of several hours. For smaller particles, the half time drops rapidly indicating that for these reaction conditions, uniform precipitates are not expected if repulsive interactions are of solely electrostatic origin.

Further evidence for non-electrostatic repulsive interactions arises from the shear stability of the particles. During the precipitation reaction particles with radii of 0.3 μm are shear aggregated. However, at the end of the reaction 1 μm particles are stable to shear aggregation. The repulsive electrostatic force grows as $\varepsilon\varepsilon_0\psi_0{}^2$ while the viscous force driving the particles together increases as $6\pi\eta\gamma a^2$. Experimentally we find that the particles grow at constant ψ_0. As a result, if a particular shear rate is capable of generating irreversible agglomerates of 0.3 μm particles (i.e., is capable of forcing particles into the primary minimum), the same shear rate should be capable of producing agglomerates of 1 μm particles. This is not observed. Shear induced agglomeration only occurs during the reaction; a result indicating that there

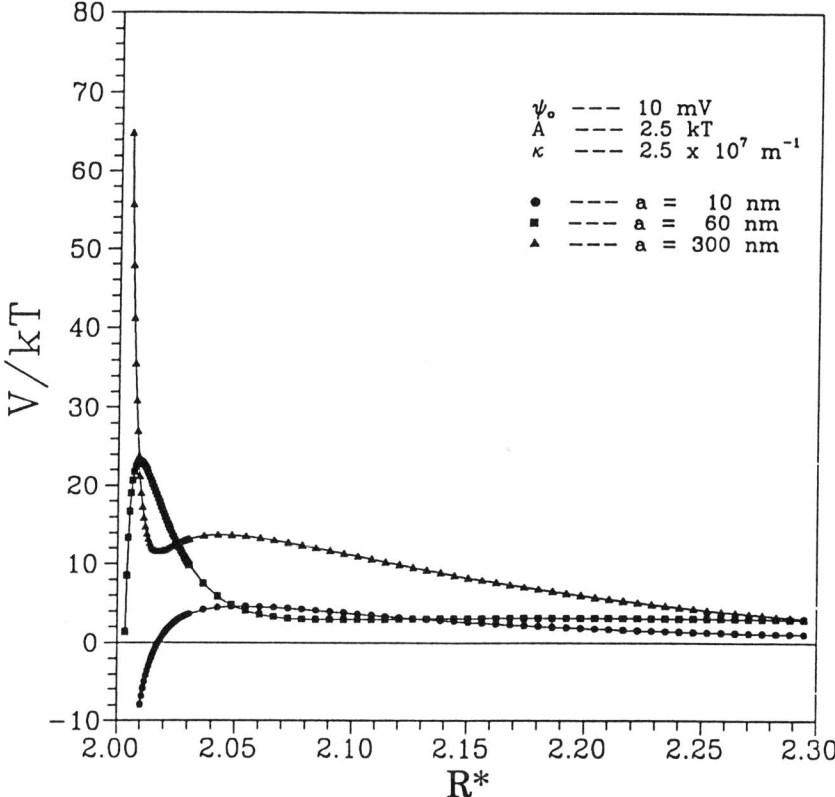

Figure 2 Total interaction potential as a function of pair separation for particles of various size.

is a repulsive interaction in addition to electrostatic forces.

We propose that the additional repulsive interaction is short range in nature and can be modelled on solvation interactions observed in other metal oxide systems.[15,16] The interaction decreases with R^* in an exponential manner with a decay length of 1 nm and has a form

$$V_s(R^*) = \pi\, S_a\, l_s\, a\, \exp[a/l_s(R^* - 2)] \tag{5}$$

where S_a takes on a value of 1.5×10^{-3} Jm^{-2}. For a variety of particle sizes fig. 2 shows pair potentials for surface potentials of 10 mV, A= 2.5 kT, and $\kappa = 2.5 \times 10^7$ m^{-1} as determined from the conductivity of the mother liquor for a reaction of 0.05 M TEOT and 0.25 M H_2O for a variety of sizes. As can be seen the pair potential develops a repulsive barrier large enough to stop Brownian aggregation for particles with radii larger than 50 nm. This barrier is due to a sum of solvation and van der Waals interactions and is called the solvation barrier. With increases in particle size, the solvation maximum grows to a level such that neither shear or Brownian motion can force particles into the primary minimum. However, electrostatic interactions become important and a secondary minimum (the solvation minimum) develops. The barrier between the solvation minimum and larger separations is largely controlled by van der Waals and electrostatic interactions and is referred to as the electrostatic maximum.

The importance of the pair potentials shown in fig. 2 can be understood by considering particle growth in a precipitation reaction. Small particles are nucleated and begin to increase in size. As indicated by fig. 2, particles smaller than approximately 50 nm are unstable and will grow by a combined mechanism of aggregation and molecular addition. Once larger than 50 nm, particle growth by aggregation with particles of the same size is greatly reduced and a constant number density of colloidally stable particles will be generated. Stability for these particles is determined by the solvation forces as electrostatic interactions are too weak to be effective in this size range. However as particles continue to grow, the electrostatic interaction becomes important and the solvation minimum develops. For electrostatic barriers which are sufficiently low, particles can diffuse into the solvation minimum and if the resulting doublets have a lifetime sufficient for precipitating material to fuse the pair together, agglomerated morphologies will result. For larger surface potentials, the electrostatic maximum at the particle size where the solvation minimum develops will increase. At $\psi_0 \geq$ 15-18 mV for the range of κ conditions studied here, the electrostatic maximum is too high for particles to diffuse over. As a result, uniform precipitates will be formed. However, the existence of the solvation minimum can be probed by the action of shearing.

The solvation maximum is sufficiently high for particles larger than 50 nm that neither Brownian motion or shear rates of reasonable magnitude are capable of driving pairs into the primary minimum. However, modest shear rates can push particles over the electrostatic maximum and form solvation minimum doublets. If these doublets have sufficient lifetime for precipitating material to fuse the pair together, agglomerated morphologies will result. On the other hand, as there is no reactive material at the end of the reaction, shearing the precipitate after the soluble titanium concentration has reached its equilibrium value will not result in irreversible agglomerates. Extensive calculations accounting for van der Waals, electrostatic, solvation interactions and viscous interactions demonstrate that for values of interaction potential parameters given above, the proposed pair potential is capable of predicting the boundary between agglomerated and uniform precipitates formed in quiescent solutions as well as the critical shear rate for the formation of agglomerated morphologies where ψ_0 is greater than 12 mV.[13]

CONCLUSIONS

These studies demonstrate that pair potentials play an important role in the development of the precipitate morphologies. By varying the magnitude of the pair potentials, morphologies can be manipulated between agglomerates and uniform particles. We conclude that the physical chemistry governing particle interactions (and thus whether precipitates form as discrete particles) can be largely decoupled from the details of the reaction chemistry. As a result, when faced with a new chemistry and searching for methods of preparing uniform precipitates, rather than focussing on the length of the nucleation period or the details of the bond making and breaking mechanisms, attention should be focussed on developing colloidal stability of the precipitating material.

ACKNOWLEDGMENTS

This work was supported by the U.S. Department of Energy through the Materials Research Laboratory at the University of Illinois through grant DOE AC0276-ER01198.

REFERENCES

1. Matijevic, E., *Acc. Chem. Res.* 14, 22 (1981)
2. Matijevic, E., Bell, A., Brace, R., and McFadyn, P., *J. Electrochem. Soc.* 120, 893 (1973)
3. Matijevic, E., *Ann. Rev. Mat. Sci.* 15, 483 (1985)
4. Barringer, E. A. and Bowen, H. K., *Langmuir* 1, 414 (1985)
5. Barringer, E. A. and Bowen, H. K., *J. Am. Cer. Soc.* 65, C-199 (1982)
6. Jean, J. H., and Ring, T. A., *Langmuir* 2, 251 (1986)
7. Jean, J. H., and Ring, T. A., *Colloids and Surfaces* 29, 273 (1988)
8. Edelson, L. H., and Glaeser, A. M., *J. Am. Cer. Soc.* 71, 225 (1988)
9. Harris, M. T., and Byers, C. H., *J. Non-Cryst. Solids* 103, 49 (1988)
10. R. J. Bailey and M. L. Mecartney, Mat. Res. Soc. Symp. Ser. 180, 153 (1990)
11. Look, J-L., Bogush, G. H., and Zukoski, C. F., *Faraday Disc.* 90, 345 (1990)
12. Zukoski, C. F., Chow, M. K., Bogush, G. H., and Look, J-L., *Mat. Res. Symp.* 180, 131 (1990)
13. Look, J-L., Ph.D. Thesis, "Formation of Uniform Particles: Morphology and Colloidal Stability, and Stable Particle Growth During Precipitation from Titanium Alkoxides," University of Illinois, 1991
14. Russel, W. B., Saville, D. A., and Schowalter, W. R., Colloidal Dispersions, Oxford University Press, Oxford, 1989
15. Israelachvilc, J. N., Intermolecular and Surface Forces, Academic Press, NY (1985).
16. Pashley, R. M., *J. Colloid Interface Sci.* 102, 23 (1984).

PRECIPITATION OF OXALATES FROM HOMOGENEOUS SOLUTION: SYNTHESIS OF BaTi(C$_2$O$_4$)$_2$ AND Ba, Y, AND Cu OXALATES

Wendell E. Rhine, Robert B. Hallock, and Michael J. Cima
Ceramics Processing Research Laboratory, Massachusetts Institute of Technology, Cambridge, MA 02139

William M. Davis
Department of Chemistry, Massachusetts Institute of Technology, Cambridge, MA 02139

ABSTRACT

Precipitation from homogeneous solution is one approach being investigated to control the chemical and physical properties of ceramic powders. The degree of supersaturation directly affects the particle nucleation processes and can be used to control the crystal size and particle morphology. Low degrees of supersaturation lead to a low number of nuclei, and regular crystal growth predominates and results in large crystals. High degrees of supersaturation lead to a large number of nuclei, resulting in monodispersed, submicrometer particles in ideal cases. This paper discusses two examples, BaTiO$_3$ and YBa$_2$Cu$_3$O$_{7-x}$, both important electronic materials which will illustrate the versatility of precipitation from homogeneous solution.

INTRODUCTION

Precipitation from homogeneous solution is one of the best methods for controlling the particle size, size distribution, stoichiometry, and crystal habit or morphology of ceramic powders. Numerous oxide particles that are uniform in size, shape, and composition can be prepared by homogeneous precipitation methods, and these reactions can be carried out in a variety of ways, as reviewed by Sugimoto[1] and Matijevic.[2]

The first step in precipitation is the formation of primary nuclei, which may be only a few ion pairs in size. The formation rate of primary nuclei is negligible until a considerable degree of supersaturation is reached. Thereafter, the formation rate of nuclei increases very quickly and is directly related to the degree of supersaturation. Supersaturated solutions of the solute are required because the supersaturated state

8

is unstable and reverts to the stable state by precipitating the excess solute from the solution. Once formed, the primary nuclei act as seeds and grow larger to become colloidal particles (0.001-0.1 μm). After colloidal particles have formed, the particle growth process may take one of two directions: colloidal particles may be stable and remain small, or they may grow to fine crystals. If given enough time, the stable colloid may agglomerate or the fine crystals may grow to larger ones.

Nucleation is the predominant process at high degrees of supersaturation, whereas regular crystal growth predominates at low degrees of supersaturation. Therefore, if large, well-formed crystals are desired, it is necessary to keep the degree of supersaturation low. Conversely, if small crystals are desired, a large number of nuclei, and consequently a high degree of supersaturation, are required. However, if the degree of supersaturation is very high, extremely small, gel-like precipitates may form and be very difficult to isolate.

Barium titanate and high T_c superconductors are two widely investigated electronic materials which can be prepared from the oxalates. In this paper we report on two different precipitation methods: one involving low degrees of supersaturation to obtain large crystals of barium titanyl oxalate and another using high degrees of supersaturation to obtain small crystals of barium, yttrium, and copper oxalates.

EXPERIMENTAL PROCEDURE

Single crystals of barium titanyl oxalate (BTO): Ammonium titanyl oxalate (0.45 g, 1.5 mmol) was dissolved in 250 mL water, and the pH was adjusted to 1 with nitric acid. The ammonium titanyl oxalate was added to a solution of barium nitrate (0.38 g, 1.5 mmol in 250 mL water adjusted to pH 1 with HNO_3), and a clear solution was obtained. The water was allowed to evaporate slowly from an open Erlenmeyer flask, and after several months crystals of $BaTiO(C_2O_4)_2 \cdot 5H_2O$ nucleated and grew. After six months, the crystals were isolated by filtration and air-dried at room temperature.

Homogeneous precipitation of Y, Ba, and Cu oxalates: Aqueous solutions of Cu, Ba, and Y acetates were prepared and diluted with *n*-propanol to give solutions which contained greater than 4:1 alcohol:water ratios. Acetic acid was added as the acid to catalyze the decomposition of diethyloxalate. These solutions were heated to 80-85°C and then diethyloxalate was added. After the reaction was complete, the particles were isolated by filtration.

RESULTS AND DISCUSSION

Synthesis of Barium Titanyl Oxalate

Precipitation of BTO has been investigated extensively as an approach for preparing barium titanate ($BaTiO_3$) powders with controlled stoichiometry. When the reaction is controlled properly, the Ba:Ti ratio of the barium titanate produced is exactly 1:1.[3] Although BTO has been the subject of more than 40 publications, it has never been structurally characterized, and its existence as a thermodynamically stable compound is the subject of debate in the current literature.[4-6]

Barium titanyl oxalate was synthesized by reacting $(NH_4)_2TiO(C_2O_4)_2 \cdot H_2O$ with barium nitrate in dilute solution. By slowly evaporating the water from these dilute solutions, the concentration of BTO gradually increased. This method maintained low degrees of supersaturation and only a small number of nuclei formed. As expected, large, well-defined crystals of BTO were isolated. X-Ray crystallography and elemental analysis indicated the BTO contained five water molecules instead of the four usually reported.

A single crystal X-ray diffraction structural determination was carried out on one of the crystals, and the compound was found to crystallize in the monoclinic space group $P2_1/n$, with cell parameters of a = 13.367(1) Å, b = 13.852(1) Å, c = 14.023(1) Å, and β = 91.61(2)°. The unit cell contains four $Ba_2Ti_2O_2(C_2O_4)_4 \cdot 10H_2O$ asymmetric units. The overall structure of BTO is rather complex and consists of hydrated Ba cations coordinated to $[TiO(C_2O_4)_2]_4^{-8}$ anions. The $[TiO(C_2O_4)_2]_4^{-8}$ groups (Fig. 1a) are eight membered rings consisting of alternating Ti-O bonds. Similar anions are found in the starting material, $(NH_4)_2TiO(C_2O_4)_2 \cdot H_2O$,[7] and other titanyl complexes.[8-11]

The two non-equivalent Ba atoms [Ba(1) and Ba(2)] are connected by three bridging oxalate oxygens, and the two equivalent Ba(2) atoms are connected by two bridging oxalate oxygens, as shown in Figure 1b. In addition to the three oxalate oxygens, there are six water molecules associated with Ba(1), making it nine coordinate. Two of these waters (O5 and O6) have shorter Ba(1)-O bond distances than the other coordinated waters, making them more difficult to remove during thermal decomposition. Barium(2) is coordinated to one water and nine oxalate oxygens making Ba(2) ten coordinate. The shortest Ba-Ti distance is 5.841(2) Å between Ba(1) and Ti(1). The other three water molecules are waters of crystallization. As expected, these waters of crystallization are easily lost at room temperature.

Precipitation of Ba, Y, and Cu Oxalates

Synthesis of YBCO powders from the oxalates has been investigated by many different groups.[12-14] For most investigators, obtaining complete precipitation of the oxalates in aqueous solution has proven difficult because of the precise control of concentration and pH required. Irrespective of the precautions taken, the ratios of the metals in the precipitated oxalates always differ from those of the initial solutions of metal salts. To overcome this problem, Yamamoto et al.[14] and Awano et al.[15] improved the recovery of oxalates by using alcohol solvents. Use of alcohol solvents reduces the solubility of the oxalates and has a tendency to increase the

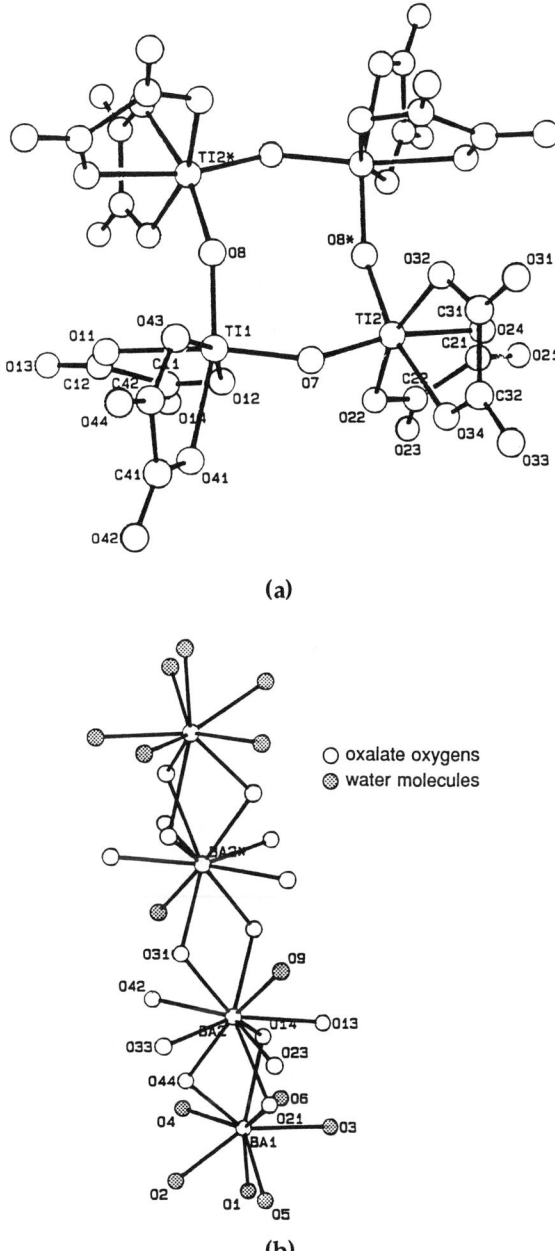

(a)

(b)

O oxalate oxygens
⊛ water molecules

Figure 1 Molecular geometry of BTO around the (a) Ti atoms and (b) Ba atoms.

Table 1 Reaction Conditions for Precipitating Ba, Y, and Cu Oxalates from Homogeneous Solution.

| Reagent (mol/L) | Reaction | | | | | |
	1	2	3	4	5	6
Y(OAc)$_3$	0.01	0.01	0.01	0.005	0.0025	0.0025
Ba(OAc)$_2$	0.02	0.02	0.02	0.01	0.005	0.005
Cu(OAc)$_2$	0.03	0.03	0.03	0.015	0.0075	0.0075
Et$_2$C$_2$O$_4$	0.07	0.14	0.27	0.27	0.27	0.54
solvent	HOAc	i-PrOH	n-PrOH	n-PrOH	n-PrOH	n-PrOH
water (mL)	200	200	200	150	100	100
HOAc		20	20	20	20	20
temp (°C)	90	68	65	75	80	80
nucleation	40 min	30 min	1 min	15 s	10 s	5 s
dispersant					HPC	HPC

HPC = hydroxypropyl cellulose (MW = 100,000).

number of nuclei formed, producing many very small particles. These small particles undergo some sintering during the conversion to YBCO and result in large, vermicular particles which are milled to obtain submicrometer particles (see Fig. 2)

We have successfully precipitated yttrium, barium, and copper oxalates from homogeneous solution by controlling the degree of supersaturation so that it is not too high throughout the process, thus avoiding the formation of gel-like precipitates. This insures formation of larger particles that are more easily isolated than with direct precipitation methods. In the method used here, oxalic acid was generated *in situ* by saponifying oxalic acid esters according to the equation

$$2H_2O + ROCOCOOR + H^+ \rightarrow H_2C_2O_4 + 2ROH.$$

The number of nuclei in this process is controlled by the rate of the reaction generating the precipitant which, in turn, is dependent on the concentration, catalyst, and temperature used for the reaction. Since the saponification of oxalic esters proceeds slowly at room temperature, the reactions were carried out at elevated temperatures (~80°C). Particles nucleated within 1 h when stoichiometric amounts of diethyl oxalate were added and within 5 s when large excess amounts of diethyloxalate were used.

The precipitation of Ba, Y, and Cu oxalates from homogeneous solution was carried out using various amounts of diethyloxalate, as shown in Table 1. During the precipitation process, each metal oxalate precipitated as its solubility product was exceeded, thus producing complete segregation. Since the yttrium oxalate was the

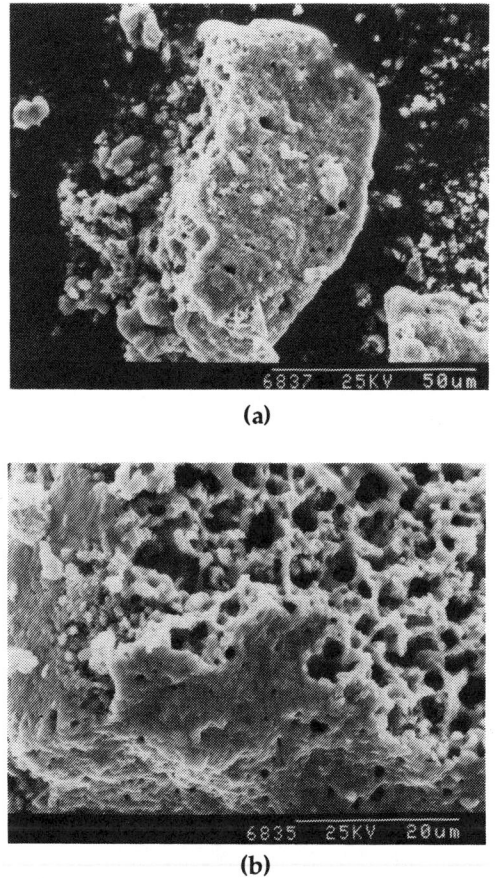

(a)

(b)

Figure 2 Particles obtained after calcining the directly precipitated oxalates at 900°C.

least soluble, it precipitated first. Then the copper oxalate precipitated and finally the barium oxalate formed. When less than fourfold excess diethyloxalate was used, the products were deficient in barium, and barium oxalate was observed to precipitate over a period of days. When greater than fourfold excess amounts of diethyloxalate were used, the reaction appeared to be complete within 3-4 h at 80°C.

The morphologies of the Ba, Y and Cu oxalates were different. In separate experiments we showed that the spherical particles observed in Figure 3a are copper oxalate. As the amount of diethyloxalate was increased, the particle size of the

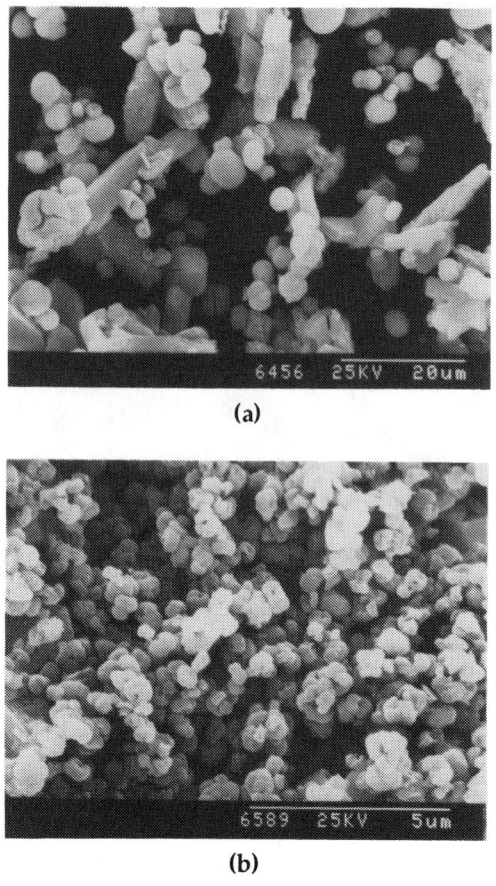

(a)

(b)

Figure 3 Particles obtained using (a) twofold excess and (b) 16-fold excess diethyloxalate.

precipitated oxalates decreased and, when 16–32-fold excess of diethyl oxalate was used, the particle sizes were ~1 µm (Fig. 3b). What is interesting is that during the thermal decomposition of the oxalates, the particles tended to maintain their morphology. After conversion to YBCO, the morphology of the original particles precipitated from reactions using twofold excess amounts of diethyloxalate could still be seen. However, the particles became porous and had the same type of morphology as the particles obtained from calcining the directly precipitated oxalates. The advantage of direct precipitation from the oxalates lies in the reduced size of the particles, which are only 2-15 µm (Fig. 4) instead of 20-50 µm in diameter

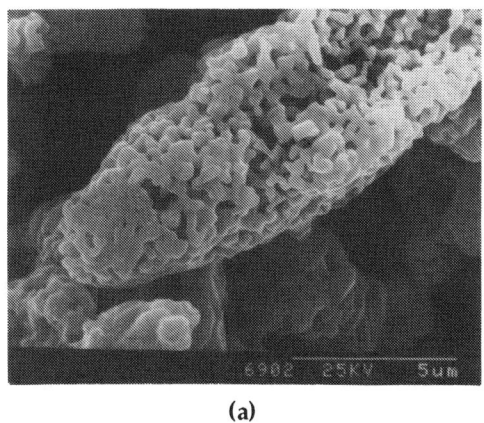

(a)

(b)

Figure 4 Particles obtained after calcining at 900°C (twofold excess).

(Fig. 2). Particles precipitated from reactions using 16–32-fold excess amounts of diethyloxalate decomposed to give micrometer sized particles of YBCO.

ACKNOWLEDGMENTS:

The authors wish to thank AFOSR (Contract No. F49620-89-C-0102) and DARPA (Contract No. MDA972-88-K-006) for financial support.

REFERENCES

1. T. Sugimoto, "Preparation of Monodispersed Colloidal Particles," *Adv. Colloid*

Interface Sci., **28**, 65-108 (1987).

2. E. Matijevic, "Monodispersed Colloids: Art and Science," *Langmuir*, **2**, 12-20 (1986).

3. W.S. Clabaugh, E.M. Swiggard, and R.J. Gilchrist, "Preparation of Barium Titanyl Oxalate Tetrahydrate for Conversion to Barium Titanate of High Purity," *J. Res. Nat. Bur. Stand.*, **56**, 289 (1956).

4. T.T. Fang and H.B. Lin, "Factors Affecting the Preparation of Barium Titanyl Oxalate Tetrahydrate," *J. Am. Ceram. Soc.*, **72**, 1899 (1989).

5. T.T. Fang, H.B. Lin, and J. B. Hwang, "Thermal Analysis of Precursors of Barium Titanate Prepared by Coprecipitation," *J. Am. Ceram. Soc.*, **73**, 3363 (1990).

6. K. Osseo-Asare, F.J. Arriagada, and J.H. Adair, "Solubility Relationships in the Coprecipitation Synthesis of Barium Titanate: Heterogeneous Equilibria in the Ba-Ti-C_2O_4-H_2O System"; p. 47 in Ceramic Transactions, Vol. 1. Edited by G. Messing, E. Fuller, Jr., and H. Hausner. American Ceramic Society, Westerville, OH, 1988.

7. G.M.H. Van de Velde, S. Harkema, and P.J. Gellings, "The Crystal and Molecular Structure of Ammonium Titanyl Oxalate," *Inorg. Chim. Acta*, **11**, 243 (1974).

8. M. Haddad and F. Brisse, "The Alkali Metal Complexes of Titanium (IV) Oxalates," *Can. Mineralogist.*, **16**, 379 (1978).

9. A.C. Skapski and P.G.H. Troughton, "The Crystal and Molecular Structure of Cyclotetra[μ-oxo-chloro-π-cyclopentadienyl-titanium(IV)]," *Acta Cryst.* **B26**, 716 (1970).

10. J.L. Petersen, "Polynuclear Oxo-Bridged Cyclopentadienyl Transition-Metal Complexes. Formation and Structural Characterization of the Titanoxane Tetramer, [$(\eta^5$-$C_5H_4CH_3)TiCl(\mu$-O)]$_4$," *Inorg. Chem.*, **19**, 181 (1980).

11. K. Wieghardt, U. Quilitzsch, J. Weiss, and B. Nuber, "Kinetics and Mechanism of Some Reactions of Chelated Complexes of Titanium (IV) with Hydrogen Peroxide. Synthesis and Crystal Structure of Cesium Teta-μ-oxo-tetra-kis[(nitrilotriacetato)titanate(IV)] Hexahydrate," *Inorg. Chem.*, **19**, 2514 (1980).

12. F.G. Sherif, Alkaline Oxalate Precipitation Process for Forming Metal Oxide Superconductors, U.S. Pat. No. 4 804 649, February 14, 1989.

13. F. Callaud, J-F. Baumard, and A. Smith, "A Model for the Preparation of $YBa_2C_3O_{7-\delta}$," *Mat. Res. Bull.*, **23**, 1273 (1988).

14. T. Yamamoto, T. Furusawa, H. Seto, K. Park, T. Hasegawa, K. Kishio, K. Kitazawa, and K. Fueki, "Processing and Microstructure of Highly Dense Barium Lanthanide Copper Oxide ($Ba_2LnCu_3O_7$) Prepared from Co-precipitated Oxalate Powder," *Supercond. Sci. Technol.*, **1**, 153 (1988).

15. M. Awano, M. Tanigawa, H. Takagi, Y. Torii, A. Tsuzuko, N. Murayama and E. Ishii, "Synthesis of Superconducting Y-Ba-Cu-O Powder by Spray Drying Method," *J. Ceram. Soc. Jpn. Inter. Ed.*, **96**, 417 (1988).

ALKOXIDE SYNTHESIS OF Al_2O_3 AND $Y_3Al_5O_{12}$ POWDERS

J. McKittrick, K. Kinsman, and S. Connell
Materials Science Program, University of California, San Diego, La Jolla, CA 92093-0411

E. Sluzky and K. Hesse
Hughes Aircraft Co., Carlsbad, CA 92008

ABSTRACT

Oxides produced by alkoxide decomposition have controllable purity and particle sizes. Alumina and yttrium aluminate amorphous powders were synthesized by hydrolysis of the isopropoxides. The phase development of the Al_2O_3 was found to follow the amorphous-θ-α phase path upon heating. YAG powders were synthesized by hydrolysis of a solution containing both isopropoxides. YAG formed at heat treatment temperatures <1200°C and contained some unreacted Y_2O_3 and θ-Al_2O_3. The particle size ranged from submicron to ~5 μm.

INTRODUCTION

High purity, fine and uniform particle size oxide powders are desirable for various structural, electronic and optical applications. One important optical application is using rare earth doped $Y_3Al_5O_{12}$ (YAG) phosphors as the luminescent material in CRT screens. When YAG is doped with terbium, it becomes brightly luminescent around 550 nm and glows green. Improving the efficiency and luminescent properties of YAG:Tb^{3+} can be achieved by producing powders with a very fine and uniform particle size.[1-3] It is also desirable to lower the solid state reaction temperature of the YAG formation, which can be achieved with using fine particle size starting powders. Gowda[4] has reported the formation of YAG powders by the sol-gel process and found the gels begin to crystallize at 810°C. We have undertaken a study to synthesize Al_2O_3 and YAG powders by alkoxide decomposition to investigate the particle morphology and size and the phase development of the starting powders.

EXPERIMENTAL TECHNIQUES

Aluminum Oxide Powder Preparation

The aluminum alkoxide was prepared by reacting approximately 0.3 grams of aluminum wire of purity 99.999% with distilled isopropyl alcohol and a small amount of mercuric chloride. The mixture was refluxed at 84°C for approximately 3.5 hours. After 3.5 hours all of the aluminum metal had dissolved but the solution had partially hydrolyzed due to inadvertent exposure to the atmosphere. The solution was fully hydrolyzed with a solution of 70 vol% ethanol and 30 vol% distilled water and a white precipitate was formed. The excess alcohol was evaporated using a rotary evaporator. The powder was rinsed once with ethanol and dried in a vacuum oven at approximately 110°C for about 2.5 hours.

Yttrium Aluminate Powder Preparation

The metal alkoxide solution was prepared by reacting approximately 0.3 grams of aluminum wire of purity 99.999% and 0.6 grams of yttrium powder of purity 99.9% with distilled isopropyl alcohol and a small amount of mercuric chloride. The mixture was refluxed at 84°C for 72 hours. After 72 hours the solution was a green-brown color and it was found that all of the aluminum metal had dissolved, but not all of the yttrium metal had dissolved. Mazdiyasni, et al.[5] determined that the reaction for yttrium is typically very slow and is difficult to drive to completion. The metal alkoxide solution was decanted from the unreacted yttrium and hydrolyzed with 10 ml of a 30 vol% distilled water/70 vol% ethanol mixture. Instead of forming a precipitate, the solution started to gel. The gel was dried overnight in a vacuum oven at approximately 110°C, then crushed with a mortar and pestle. The resultant powder was a light green color.

Powder Characterization

The thermal characteristics were examined by differential thermal analysis[*] (DTA) by heating at 10°C/min in a flowing air atmosphere. Weight loss was determined by thermogravimetric analysis[*] (TGA). The surface area was measured using a BET analyzer[†] with nitrogen as the adsorbate. The surface area measurements were made on the powder which had been dried at 110°C and on a sample which had been heat treated to 1200°C. X-ray diffraction (XRD) analysis was performed on a X-ray diffractometer on the as-processed powders and heat treated samples. The morphology of the as-synthesized and heat treated powders were examined by scanning electron microscope (SEM).

---

[*]   Perkin-Elmer Corporation, Norwalk, CT.
[†]   Quantasorb Jr., Quantachrome Corp., New York.

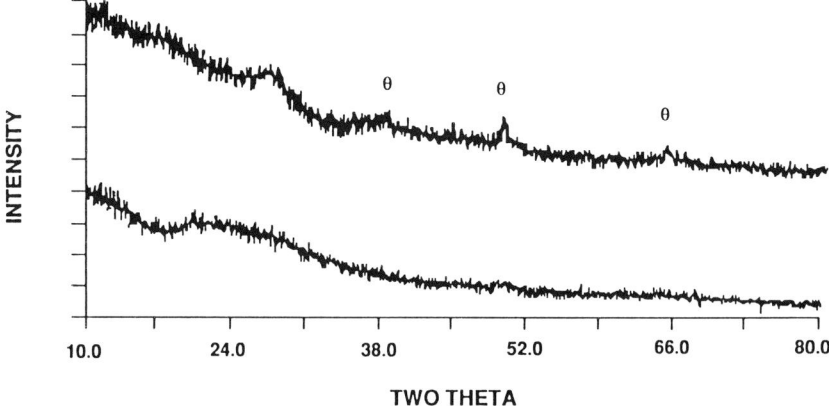

Figure 1 X-ray diffraction pattern taken on the alumina precursor powder. Bottom trace is from as-synthesized powder. Top trace is from the sample heated to 910°C. The broad peaks index to θ-Al_2O_3.

EXPERIMENTAL RESULTS AND DISCUSSION

The bottom trace in Figure 1 shows the x-ray diffraction data for the as-processed alumina precursor compound. The sample is clearly amorphous with no evidence of crystalline peaks. The TGA results indicate there are three decompositional regions, as shown in Figure 2: a weight loss of 14% from 20-167°C, 17% from 167-480°C and 5% from 480-1200°C. The decomposition of the powder starts with the evolution of residual alcohol followed by water and OH⁻ groups. At higher temperatures, persistent carbonaceous species are finally oxidized. BET results indicate that the as-synthesized powder has an average surface area of 295 m^2/g. The top trace in Figure 1 shows the sample after a heat treatment to 910°C. The majority of the sample is amorphous but some peaks from the monoclinic θ-Al_2O_3 phase are evident, which begin to appear after heat treating to 465°C. These results are in agreement with what has been reported by other researchers.[6-8] After heating to 1200°C the x-ray diffraction trace shows the presence of θ-Al_2O_3 along with some weak α-Al_2O_3 peaks. The nucleation of α-Al_2O_3 occurs at a slightly depressed temperature than what had been previously reported.[7] Further heating to 1450°C shows only α-Al_2O_3.

The analysis of the YAG precursor powders show similar behavior to the alumina precursors. Figure 3 is the DTA result on the as-processed YAG powder. A low temperature peak which occurs at 412°C is due to volatilization of hydrous compounds but no peaks are observed above that temperature. This indicates that the crystallization of YAG from the amorphous precursors occurs adiabatically or with

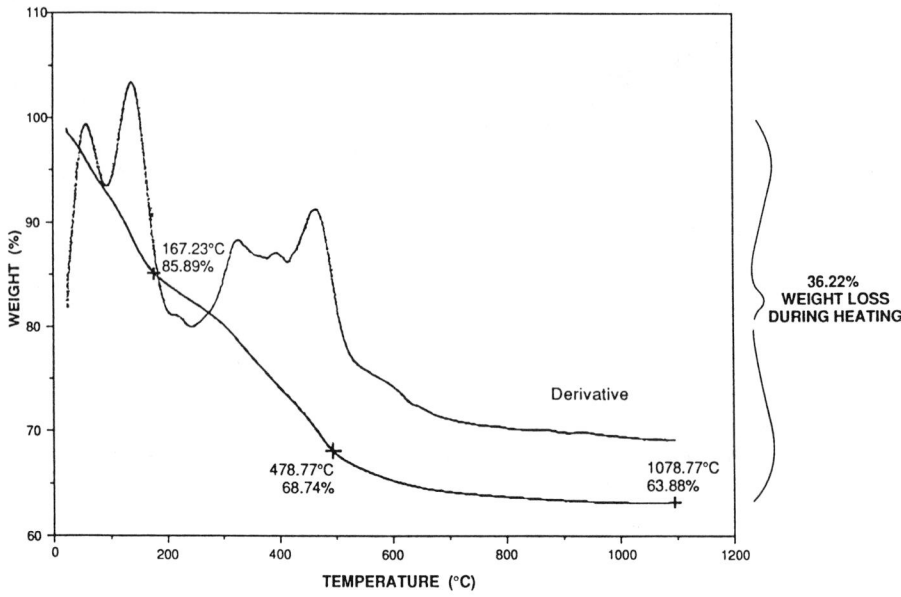

Figure 2 Thermogravimetric analysis trace taken on the Al_2O_3 precursor powder.

a very small evolution of heat. The x-ray diffraction pattern taken after heating the powder to 1200°C is shown in Figure 4. The peaks are identified as YAG with a small amount of unreacted Y_2O_3 and θ-Al_2O_3. We had an excess of aluminum metal dissolved in the original solution and expected to observe some Al_2O_3 in the final product. Finding unreacted Y_2O_3 in the powder indicates that the solid state reaction to form YAG from the individual oxides is slow. Abell, et al.[9] found that the solid state reaction between Al_2O_3 and Y_2O_3 to form YAG occurs slowly between 1300-1800°C. The presence of YAG and Y_2O_3 apparently has no influence on the nucleation kinetics of α-Al_2O_3. Other researchers have reported lowering the θ-α transformation temperature by additions of CuO and Fe_2O_3,[9] Cr and Fe,[11] but no influence was found from MgO and Cr_2O_3 additions.[12] The sample was ground and examined in the SEM for particle size analysis. Figure 5 is the SEM micrograph of the powder which shows that the particle size remains small but is broad, ranging from submicron size to >5 μm.

CONCLUSIONS

We have formed Al_2O_3 and $Y_3Al_5O_{12}$ by hydrolysis of the isopropoxides. Both oxides form powders which have extremely fine particles sizes and are amorphous. YAG can be formed by coreaction of the aluminum and yttrium metals in isopropyl

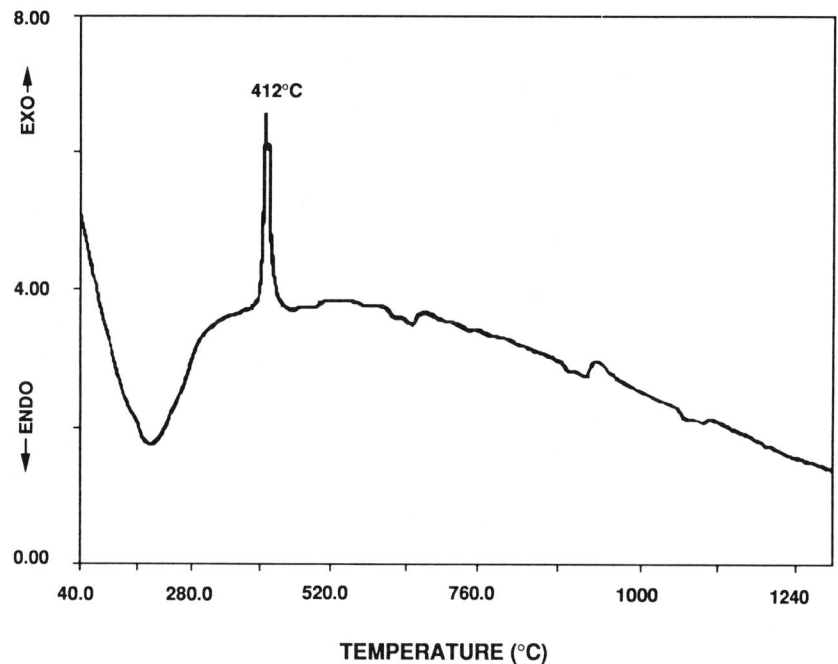

Figure 3 Differential thermal analysis trace taken of the YAG precursor powder showing only one exothermic peak at 412°C.

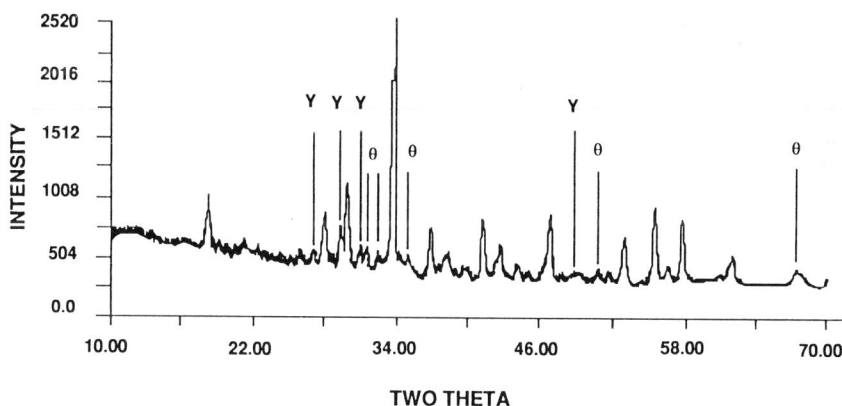

Figure 4 X-ray diffraction pattern taken from the YAG precursor powder after heat treatment to 1200°C for one hour. All peaks index to YAG except the ones labeled as $Y=Y_2O_3$ and $\theta=\theta\text{-}Al_2O_3$.

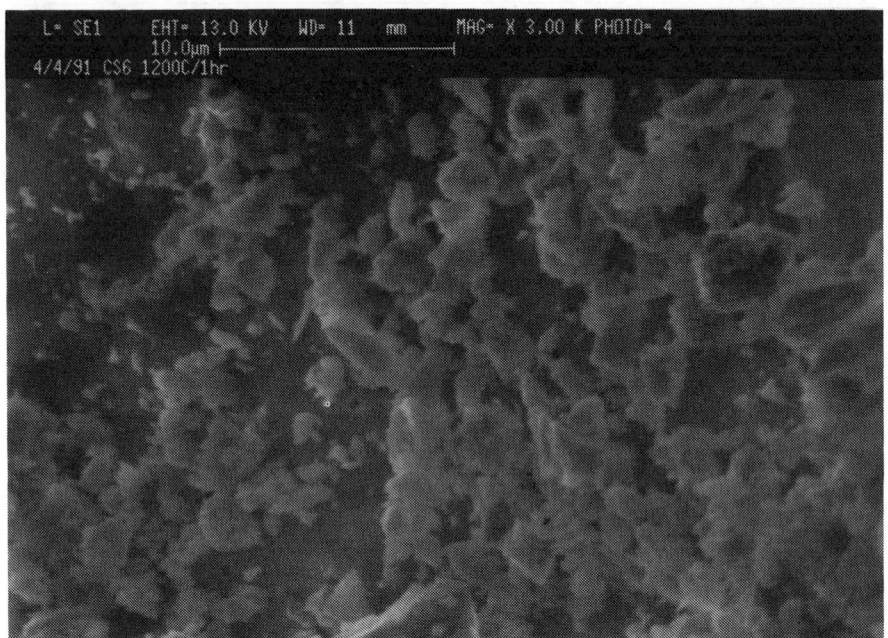

Figure 5 Scanning electron micrograph of the YAG powder after a 1200°C heat treatment.

alcohol. Crystallization of the Al_2O_3 powder follows the amorphous-θ-α path in the pure Al_2O_3. Crystallization of the YAG occurs below 1200°C with unreacted θ-Al_2O_3 and Y_2O_3 still present, indicating that the θ-α transformation temperature is not lowered by the presence of YAG and Y_2O_3.

ACKNOWLEDGMENTS

This work was supported by Hughes Aircraft Co. in Carlsbad, CA and the authors gratefully acknowledge their support.

REFERENCES

1. W. Espe, in Materials of High Vacuum Technology, Vol. 3, Pergamon Press, New York, 1968
2. S. Woodcock and J.D. Leyland, "The Choice of Phosphor for Modern CRT Display Applications", Displays, 1/2, 69-82 (July, 1979)
3. T. Takamori and L. David, "Controlled Nucleation for Hydrothermal Growth of Yttrium-Aluminum Garnet Powders", Am. Ceram. Soc. Bull., 65 [9] 1282-86 (1986)

4. G. Gowda, "Synthesis of Yttrium Aluminates by the Sol-Gel Process", Mat. Sci. Lett., 5[10] 1029-32 (1986)

5. K. S. Mazdiyasni, C.T. Lynch and J.S. Smith, "The Preparation and Some Properties of Yttrium, Dysprosium and Ytterbium Alkoxides", Inorg. Chem, 5[3] (1966)

6. M. Kumagai and G. L. Messing, "Controlled Transformation and Sintering of a Boehmite Sol-Gel by α-Al_2O_3 Seeding", J. Am. Ceram. Soc.,68[9]500-505 (1985)

7. S. J. Wilson and G.D. McConnell, "A Kinetic Study of the System Gamma AlOOH/Al_2O_3", J. Solid State Chem., 34, 315-22 (1980)

8. H.C. Stumpf, A.S. Russell, J.W. Newsome, and C.M. Tucker, "Thermal

9. J.S. Abell, I.R. Harris, B. Cockayne, B. Lent, "An Investigation of Phase Stability in the Y_2O_3-Al_2O_3 System", J. Mater. Sci, 9, 527-537 (1974)

10. Y. Wakao and T. Hibino, "Effects of Metallic Oxides on Alpha Transformation of Alumina", Nagoya Kogyo Gijutsu Shikenski Hokoku, 11, 588-95 (1962)

11. G.C. Bye and G. T. Simpkin, "Influence of Cr and Fe on Formation of α-Al_2O_3", J. Am. Ceram. Soc., 57[8]367-71 (1974)

12. F.W. Dynes and J. W. Halloran, "Alpha Alumina Formation in Alum-Derived Gamma Alumina", J. Am. Ceram. Soc., 65[9]442-48 (1982)

THE EFFECT OF YTTRIA STABILIZER
ON THE ELECTROKINETIC BEHAVIOUR
OF ZrO_2-Al_2O_3 COLLOIDAL SUSPENSIONS

D. Goski and J.C.T. Kwak
Department of Chemistry, Dalhousie University, Halifax, Nova Scotia,
B3H 4J3 (Canada)

K.J. Konsztowicz
National Research Council of Canada, Halifax, Nova Scotia, B3H 3Z1 (Canada)

ABSTRACT

Colloidal suspensions of Al_2O_3-ZrO_2 mixtures have been prepared with and without 3 mol% Y_2O_3 as stabilizer of tetragonal ZrO_2 (t-ZrO_2). The electrophoretic mobilities (EM) of oxide dispersions examined in aqueous media were compared to those obtained in 0.01M KNO_3 for compositions containing from 5 to 30 vol% ZrO_2. It was found that the points of zero charge (pzc) of the composite suspensions were displaced relative to pure Al_2O_3 due to the presence of ZrO_2. Additionally, a strong effect of the presence of yttria stabilizer on the pzc of Al_2O_3-ZrO_2 suspensions has been established.

INTRODUCTION

Previous work by the same authors[1] has confirmed the effects of the addition of colloidal ZrO_2 to Al_2O_3 dispersions on the pzc displacement of Al_2O_3-ZrO_2 composite systems in comparison to the electrokinetic behaviour of pure matrix material, Al_2O_3. It was found that this effect is valid both in water and electrolyte. Introduction of Y_2O_3 stabilizer in many types of zirconia-toughened alumina (ZTA) composites is common for the purpose of stabilization of ZrO_2 in its tetragonal form.[2] The aim of this work is to compare the effect of the presence of Y_2O_3 on the electrokinetic behaviour of Al_2O_3-ZrO_2 composite suspensions in water, which is the most common processing medium, and in constant ionic strength conditions (0.01M KNO_3).

MATERIALS

Commercial powders of Al_2O_3[*] (free of MgO), $t-ZrO_2$[†] and $t-ZrO_2$ with 3 mol% Y_2O_3[‡] (3-YSZ) were centrifugally classified[§] to submicron size. Concentrated suspensions (50.5 vol%) of these powders were prepared in distilled water at pH 4.0. Slurries were mixed to give compositions with 5, 10, 20 and 30 vol% ZrO_2 (or 3-YSZ) in Al_2O_3 with respect to total solids content. The pH values of the composite suspensions were adjusted to 4.0 with 1 N HNO_3,[¶] eqilibrated for 12 hr on a planetary mixer[**] and ultrasonically homogenized[††] for 8 min at 300 W. Samples of 6 μl taken from each dense slurry were redispersed in distilled H_2O to a concentration of 300 ppm. $NH_3(aq)$[‡‡] or HNO_3 was used to adjust the pH as necessary to give the range of selected values between 3 and 9.

TECHNIQUE

The 300 ppm suspensions were ultrasonically agitated at 20 W for 20 min. to redisperse soft agglomerates to a comparable degree. Using microelectrophoresis,[§§] the ζ-potentials of individual suspensions of Al_2O_3, ZrO_2 and 3-YSZ, as well as their above described mixtures were calculated from their electrophoretic mobilities (EM) using the Smoluchowski equation. Twenty measurements in each direction at each stationary plane of a flat cell were taken for each colloidal composition to give a sum of 80 measurements per sample. The average EM and standard deviations for each mixture and each pH were calculated from these 80 measurements.

RESULTS AND DISCUSSION

Figure 1a shows the ζ-potential versus pH for Al_2O_3, ZrO_2 and 3-YSZ in aqueous media. The pzc of ZrO_2 at pH 6.3 and Al_2O_3 at pH 8.3 are congruous with the published results of other authors.[3,4] The pzc of 3-YSZ at pH 7.2 is remarkably higher than that of pure ZrO_2. The upward shift of 0.9 pH units is not surprising bearing in mind the high pzc value for Y_2O_3 at pH 8.95 as reported by Parks.[5] Other works[6,7] have also shown that the pH_{pzc} of ZrO_2 can be increased from 0.9 to 1.2 units upon addition of 2 or 3 mol% Y_2O_3.

The ζ-potential values of pure ZrO_2, 3-YSZ and Al_2O_3 dispersed in 0.01M KNO_3 are presented in Fig. 1b. Unlike the pzc values for Al_2O_3 and pure ZrO_2, the pzc of

---

*   Reynolds RCHP-DBM , spec. surf. area=6.9 m^2/g, d_{50}=0.39μm.
†   TZ-0, Toyo-Soda (Tosoh), spec. surf. area=13.2 m^2/g, d_{50}=0.19μm.
‡   Toyo-Soda (Tosoh), spec. surf. area=13.2 m^2/g, d_{50}=0.19μm.
§   Continuous Flow Centrifuge IEC model CU 5000.
¶   ACS Reagent Grade, Fisher Scientific.
** Turbula, model 2TC.
†† Heat Systems, model W-385.
‡‡ ACS Reagent Grade, Caledon Lab. Ltd.
§§ Rank II, flat cell with white light illumination.

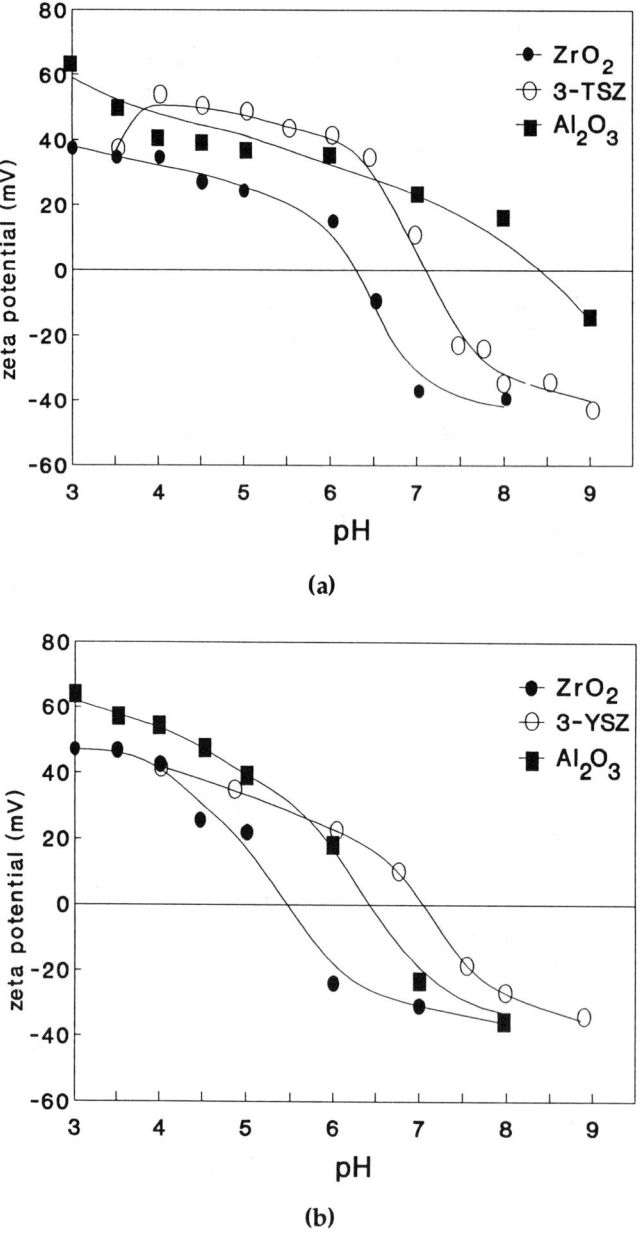

Figure 1 Microelectrophoresis results of Al_2O_3, ZrO_2 and ZrO_2 + 3 mol% Y_2O_3 dispersed in (a) H_2O and (b) 0.01M KNO_3.

3-YSZ does not differ from that in water. Simon[7] observed that Y_2O_3 is relatively soluble when in aqueous media, as compared to other oxides such as pure ZrO_2. He suggested that since the solubility of Y_2O_3 increases significantly in low pH suspensions it affects the overall proton consumption by the surface of ZrO_2 particles. Kolarik and Kourim[8] have also observed that Y^{3+} ions can adsorb on a positively charged hydroxide. This would imply that in the case of the present work it may be possible for Y^{3+} ions to readsorb on the positively charged ZrO_2 particle surface to some extent at low pH.

The comparison of results shown in Figs. 1a and 1b indicates that preparation of 3-YSZ dispersions in electrolyte involves a compression of the electrical double layer, as reflected by the decreased ζ-potential values while the pzc remains unchanged. On the other hand, the strong downward shift of pzc values measured for pure ZrO_2 and Al_2O_3 in 0.01M KNO_3 as compared to aqueous media, may indicate that specific adsorption of anions has occurred on surfaces of these oxides.[9] Given that NO_3^- ions are not assumed to specifically adsorb on the examined oxide surfaces in the range of concentrations of electrolyte,[10] in order to reexamine this possibility, the powders were washed 3 times in distilled water and then redispersed in their appropriate aqueous media. No change in the pzc was encountered, possibly meaning that if surface impurities were present in commercial powders used, they could not be removed simply by washing and that they might be responsible for generation of an affinity for the NO_3^- ions.

Figure 2a shows the data for Al_2O_3-3-YSZ dispersions in water. The electrokinetic behaviour of these mixtures appear not to be very sensitive to changes of the ZrO_2 content in the suspensions of the examined compositions. Similarly to results obtained for pure ZrO_2-Al_2O_3 described elsewhere,[1] the pzc values are intermediate of the pure components when the ZrO_2 contents is increased from 5-30 vol% ZrO_2 (with respect to total solids content). While in the case of mixtures of Al_2O_3 with pure ZrO_2 the pzc values were spread over an appreciable pH range, in the Y_2O_3 doped compositions the pzc of all mixtures remain fairly close to one another. This would indicate that these pzc values are highly affected by the presence of Y_2O_3 dopant probably because of its own high pzc value.[5] It is possible therefore that the addition of Y_2O_3 has its own strongly pronounced flocculating effect.

As indicated in Fig. 2b, the trends in electrokinetic behaviour of Al_2O_3-3-YSZ mixtures dispersed in 0.01M KNO_3 are comparable to the above described behaviour in aqueous media, although some pzc displacement for all compositions can be noticed, as well as the decrease of ζ-potential values determined for these compositions.

The comparison of figures 1a and b and figures 2a and b shows that in the lower pH region, the ζ-potential values for Al_2O_3-ZrO_2 mixtures are more widely spread (up to 12 mV) with a more obvious effect of ZrO_2 content than in Al_2O_3-3-YSZ

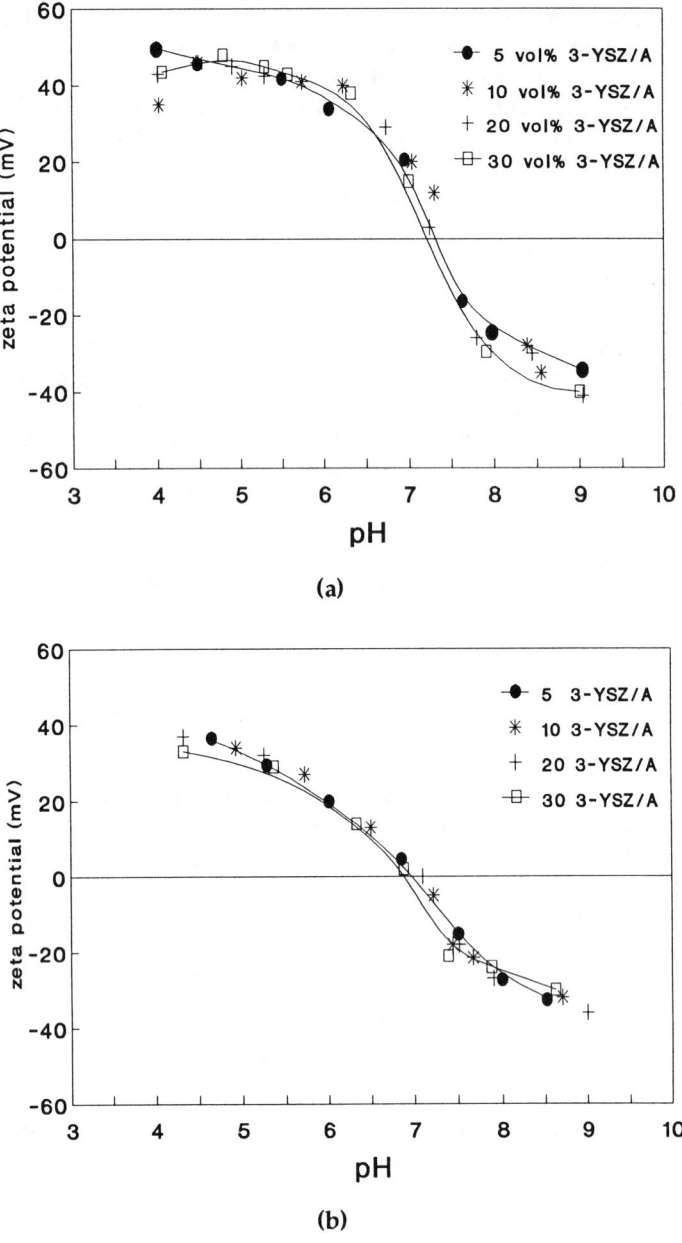

Figure 2 ζ-potential values of mixtures containing Al_2O_3 and 5-30 vol% of 3-YSZ in (a) H_2O and (b) 0.01M KNO_3.

systems.

CONCLUSIONS

The pzc of 3-YSZ is unaffected by the type of measuring medium (0.01M KNO_3 as compared to water). Possibly due to solubility of Y_2O_3, the NO_3^- anion is unable to adsorb onto the particle's surface and remains indifferent.

In Al_2O_3-3-YSZ composite suspensions, little effect of ZrO_2 content on composite pzc was found. The presence of 3 mol% addition of Y_2O_3 seems to dominate over the effect of increasing ZrO_2 content (in the range of examined compositions with 5-30 vol% ZrO_2).

The lack of obvious modifications in ζ-potential values of Al_2O_3-3-YSZ composite suspensions in the examined pH range, as compared to Al_2O_3-pure ZrO_2 systems, may be caused by a strong flocculating effect brought about by the presence of Y_2O_3 dopant.

Issued under NRCC No. 32974.

REFERENCES

1. D. Goski, J.C.T. Kwak and K. Konsztowicz,"Electrokinetic Behaviour of Zirconia-Alumina Colloidal Suspensions in Water and in Electrolyte." Cer. Eng. and Sci. Proc. 1991, to be published.
2. N. Claussen,"Microstructural Design of Zirconia-Toughened Ceramics (ZTC)." pp.325-351,in Advances in Ceramics, 12, pp.325-351, eds. N.Claussen, M.Rühle and A.Heuer, The Am. Ceram. Soc., OH, 1984.
3. R. Moreno, J. Requena and J.S. Moya, "Slip Casting of Yttria Stabilized Tetragonal Zirconia Polycrystals," J. Am. Ceram. Soc., 71 [12] 1036-1040 (1988).
4. D.C. Agrawal, R. Raj and C. Cohen, "Nucleation of Flocs in Dilute Colloidal Suspensions", J. Am. Ceram. Soc., 72 [11] 2148-2153 (1989).
5. G.A. Parks, "The Isoelectric Points of Solid Oxides, Solid Hydroxides and Aqueous Hydroxo Complexe Systems," Chem. Rev., 65 177-196 (1965).
6. E.M. DeLiso, W.R. Cannon and A.S. Rao,"Dispersions of Alumina-Zirconia Powder Suspensions," pp.335-341, in Advances in Ceramics, 24, The Am. Ceram. Soc.,OH, 1988.
7. C. Simon,"Characterization and Surface Properties of Yttria Stabilized Zirconia." To be published in proceedings of VIIth CIMTEC (World Ceramic Congress), ed. P.Vincenzini, Elsevier Pub., London.
8. Z. Kolarik and V. Kourim, "Radioactive Isotope Adsorption on Precipitates IV. Adsorption of Yttria on Fe(OH)$_3$," Collect. Czech. Chem. Comm., 26 1082-1091 (1961).
9. J. Lyklema, "Fundamentals of Electrical Double Layers in Colloidal Systems," pp.47-70, in Colloidal Dispersions, ed. J.W.Goodwin, Royal Soc. of Chem.,

London, 1981.

10. J.A. Schwarz, C.T. Driscoll and A.K. Bhanot, "The Zero Point of Charge of Silica-Alumina Oxide Suspensions," J. Coll. Interface Sci., 97 [1] 55-61 (1984).

A SURFACE CHEMICAL TECHNIQUE FOR SINTERING AID ADDITION

S. G. Malghan and P. S. Wang
Ceramics Division, National Institute of Standards and Technology,
Gaithersburg, MD 20899

A. Sivakumar
Nalco Chemical Company, Chicago, IL

ABSTRACT

A homogeneous distribution of sintering aids is essential in the preparation of silicon nitride slurries. A novel technique to homogeneously distribute alumina on silicon nitride particles was developed based on electrostatic attractive forces between dissimilar particles. In addition, applicability of the measurement of acoustophoretic mobility of suspensions was demonstrated for the study of coating of alumina on the silicon nitride particulates. The alumina coating was characterized by surface sensitive x-ray photoelectron spectroscopy (XPS) technique. Particle size distribution of alumina was identified as one of the major parameters affecting the coating on silicon nitride particles.

HOMOGENEITY OF SINTERING AID DISTRIBUTION

Uniform distribution of sintering aid in the processing of silicon nitride powder provides enhanced properties to the resulting ceramic. Especially, pressureless sintering of Si_3N_4 powder is possible in the presence of liquid-forming sintering aids that promote the sintering process by dissolution, diffusion and reprecipitation of Si_3N_4.[1] The microstructure of the dense ceramic, in general, is controlled by the ability of oxide mixtures to form a uniform grain boundary phase. Inhomogeneities (agglomerates of single or mixtures of oxides and concentration gradients) introduced due to sintering aid distribution may form voids, chemical gradients and other flaws during densification.

Sintering aids addition is usually carried out by mechanical mixing, milling, or chemical techniques. Chemical routes appear to provide enhanced homogeneity.[1,2] However, chemical techniques suffer from various problems, such as high cost of raw materials and processing, formation of hydroxides, and difficulty of process control at high solids concentration. The following are some examples of the chemical techniques:

1) precipitation of hydroxides from[1,2]

$$Y(NO_3)_3 \xrightarrow{\;OH^-\;} Y(OH)_3,$$

—inorganic salts,

$$Al(OC_4H_9)_3 \xrightarrow[heat]{\;H_2O\;} Al(OH)_3$$

—alkoxides,

2) surface interactions
   —organometallic surfactants,[3]
   —electrostatic and electrosteric attraction,[4]

In the present study, the addition of alumina as a sintering aid for Si_3N_4 was examined by the application of electrostatic attraction as a driving force. In addition, the coating effectiveness was characterized by acoustophoretic mobility and XPS.

EXPERIMENTAL

As a model system, SN E-03 (d_{50} = 1.1 μm and surface area = 3.1 m^2/g) Si_3N_4 powder from Ube Industries*, and two alumina powders of different size distribution were used. AKP-50,(A), from Sumitomo Electric Co., had a d_{50} = 0.2 μm and a surface area of 10.6 m^2/g; whereas, Dispal-180,(B), from Vista Chemical Co., had a d_{50} = 40 nm and a surface area of 180 m^2/g. The procedure consisted of ultrasonication of Si_3N_4 slurry (2% by volume in distilled water), adjustment of pH by using NH_4OH or HNO_3, and stirring for 30 min. to equilibrate the surfaces. A slurry of alumina powder (0.3% by V) was also prepared by the same procedure. The two suspensions were mixed and stirred for additional 30 min. The resulting suspensions were evaluated by acoustophoresis, and dry powders (100°C) by XPS. All acoustophoretic mobility measurements were carried out without the addition of a supporting electrolyte at a solids concentration of 2% by volume for Si_3N_4 and 0.3% by volume for Al_2O_3. The XPS samples were prepared by uniformly sprinkling on a sticky tape. A polyacrylic acid (PAA) of 5000 molecular weight from Aldrich Chemicals was used as a dispersant.

---

* Certain trade names and company products are mentioned in text in order to adequately specify the experimental procedure and equipment used. In no case does such identification imply recommendation or endorsement by National Institute of Standards and Technology, nor does it imply that the products are necessarily the best available for the purpose.

RESULTS AND DISCUSSION

Only two methods of surface charge modification were examined for producing oppositely charged Si_3N_4 and Al_2O_3 particles. The most straight forward technique is the pretreatment of both powders at an appropriate pH at which the two powders carry opposite charges. The second method is that of using a polyelectrolyte to enhance the negative charge on Si_3N_4 powder surface to carry out controlled coagulation of negatively charged Si_3N_4 and positively charged Al_2O_3 particles.

The pH dependence of acoustophoretic mobility (electrokinetic sonic amplitude, ESA) as a function of suspension pH for Si_3N_4, Si_3N_4 containing 100 ppm PAA, and Al_2O_3-A and B powders is presented in Fig. 1. As expected, the pH_{iep} (the pH at which the particles carry a net zero charge) of Si_3N_4 and that of Si_3N_4-containing-100 ppm PAA are at 5.1 and 2.9, respectively. The pH_{iep} of PAA alone, as measured by ESA measurement, was found to be at pH = 2.2-2.4. Therefore, the Si_3N_4-containing-100 ppm PAA data indicate that the surface behavior of Si_3N_4 was dominated by PAA. The data of Al_2O_3 powders ESA vs pH indicate that both powders have the same pH_{iep} (8.5 - 8.7). Though the two powders are synthesized by different chemical routes, their surface species appear to be the same. However, the ESA of Al_2O_3 powders are different at low pH due to the influence of particle concentration and dissolved species.

The ESA response of a series of Si_3N_4 powder samples, surface-reacted with Al_2O_3-A and B powders, is shown in Table 1. The Al_2O_3 concentration was varied in four

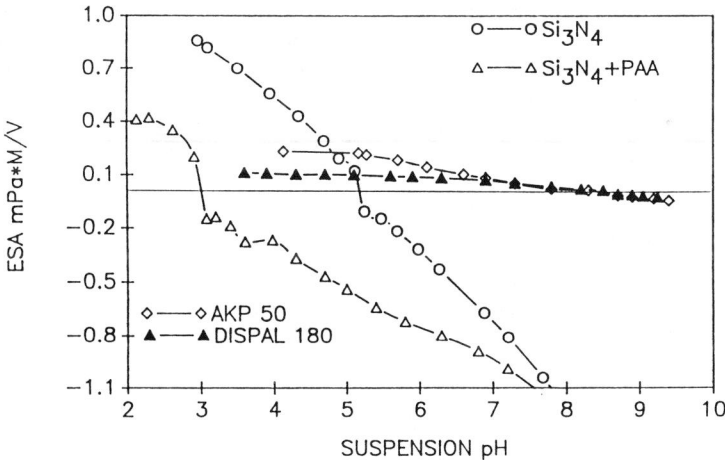

Figure 1 Change in ESA of AKP-50 (0.3% V), DISPAL 180 (0.3% V), Si_3N_4 (2% V) and Si_3N_4-containing 100 ppm PAA (2% V) as a function of suspension pH.

33

Table 1 ESA of Si_3N_4 Suspensions (2% V) in the Presence of Various Concentrations of Al_2O_3-A or B Powders Containing no PAA at pH = 5.6 ± 0.1 and 100 ppm PAA at pH = 3.8 ± 0.2.

Si_3N_4, %Wt	Al_2O_3, %Wt	No PAA ESA, mPa·M/V Powder A	Powder B	100 ppm PAA ESA, mPa·M/V Powder A	Powder B
100	0	-0.80	-0.74	-1.56	-1.58
98.5	1.5	-0.79	+0.37	-1.54	-0.35
97	3.0	-0.79	+0.76	-1.53	+0.07
95.5	4.5	-0.73	+0.82	-1.51	+0.39
94	6.0	-0.60	+0.79	-1.51	+0.58

steps of 1.5, 3.0, 4.5, and 6.0% by weight of Si_3N_4 powder in the suspension. The coating process was carried out at pH = 5.7 ± 0.1. The ESA values of Si_3N_4 shown in Table 1 and Fig. 1 do not agree because of the differences in the experimental procedure. The ESA values in Fig. 1 are affected by the powder undergoing a series of pH conditions during titration; whereas, the ESA values reported in Table 1 (100% Si_3N_4) are conditioned only at pH = 5.7. The presence of different ionic species is contributing to the ESA values. For powder A, the negative sign of ESA did not change as higher concentration of Al_2O_3 was introduced to the suspension. On the contrary, for powder B, the ESA not only transformed to a net positive value after the first addition of 1.5% Al_2O_3, but also its magnitude increased up to 4.5% by weight of Al_2O_3. This difference in the behavior of Si_3N_4 in the presence of two types of Al_2O_3 can be explained by the surface coverage of Al_2O_3. The powder B, containing 40 nm particles, may have adsorbed on the surface of Si_3N_4 particles and thus the surface may have been made to look like that of Al_2O_3. The fineness of the particles may be one of the major reasons for their adsorption on the Si_3N_4 particles. As the particles become smaller, the ratio of charge to weight (charge concentration) and diffusion coefficient are expected to become larger, thus promoting their transport to the Si_3N_4 powder surface. The Al_2O_3-A powder appears to show very little affinity to the surface of Si_3N_4, even though the initial driving force (difference in the surface charge) between the two types of particles is the same as that between Si_3N_4 and Al_2O_3-B.

Si_3N_4 slurries, containing specific concentrations of Al_2O_3-A and B, were titrated to determine their response to the ESA as a function of pH. These data are shown in Fig. 2 (for Al_2O_3-B) and Fig. 3 (for Al_2O_3-A). The pH$_{iep}$ data in Fig. 2 confirm that Al_2O_3-B has formed a coating on the surface of Si_3N_4. The pH$_{iep}$ of Si_3N_4 increased as a function of Al_2O_3 concentration. However, no significant increase in the pH$_{iep}$

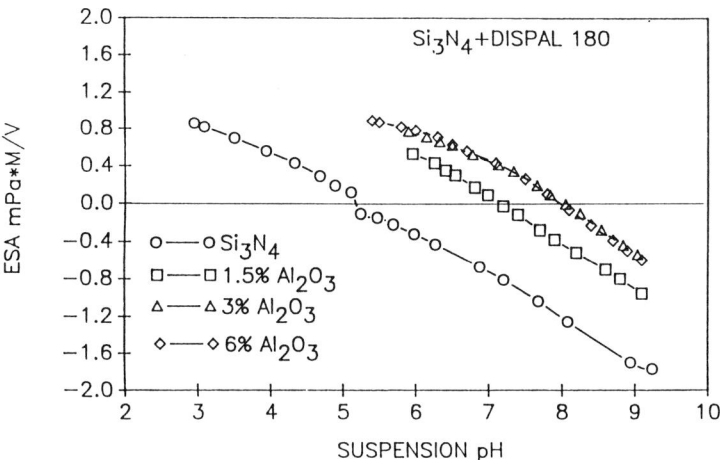

Figure 2 Comparison of ESA vs suspension pH for Si_3N_4 and Si_3N_4-containing various concentrations of Al_2O_3-DISPAL 180.

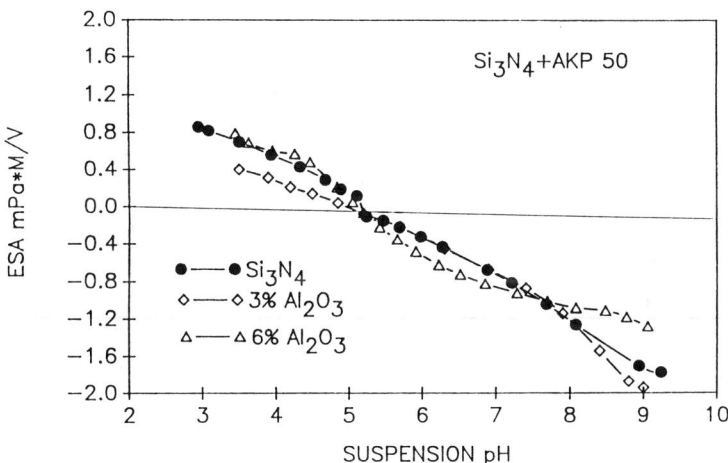

Figure 3 Comparison of ESA vs suspension pH for Si_3N_4 and Si_3N_4-containing various concentrations of Al_2O_3-AKP 50.

was observed after 3% Al_2O_3-B addition. It appears that Si_3N_4 surface is saturated and/or driving force for adsorption is too low to induce attachment of additional Al_2O_3 particles. The determination of ESA vs pH for Si_3N_4/Al_2O_3-A system in Fig. 3 showed no change of pH_{iep}, which is an evidence of no interaction between the

Table 2 Approximate Surface Compositions by XPS of Si_3N_4 and Si_3N_4/Al_2O_3-A or B Powders.

Powder	Composition, % atomic				
	C	O	N	Si	Al
Si_3N_4	29.7	23.3	24.5	22.5	—
Si_3N_4/Al_2O_3-A	43.9	25.4	12.8	14.6	3.2
Si_3N_4/Al_2O_3-B	42.1	35.0	5.4	6.2	11.3
$Si_3N_4/PAA/Al_2O_3$-B	36.6	32.0	10.3	11.0	8.2

two types of particles. For clarity, only 3% and 6% by weight Al_2O_3-A data are shown.

Surface charge on Si_3N_4 was expected to be enhanced by the addition of 100 ppm PAA to the suspension, as shown in Fig. 1. To this suspension, increasing concentrations of Al_2O_3-A or B were added. Since the surface charge (as deduced from ESA) is enhanced by the adsorption of PAA, more Al_2O_3 is expected to form a coating on Si_3N_4. According to the data of Table 1, the presence of PAA had no effect on the Al_2O_3-A coating on Si_3N_4, since no change in ESA was observed. Higher values of ESA, compared to those of containing no PAA, are indicative of higher surface charge on the particles. For Al_2O_3-B powder, a weak shift from negative to positive ESA took place at 3.0% Al_2O_3; whereas, in the absence of PAA a strong shift took place at 1.5% Al_2O_3. This difference is an indication of the presence of enhanced negative charges on the Si_3N_4 surface due to PAA adsorption. However, no clear indication exists that PAA has contributed to increased surface charges for the Si_3N_4/Al_2O_3-B system. This lack of clear evidence is probably related to the presence of unadsorbed PAA present in the Si_3N_4 suspension that was free to adsorb on Al_2O_3.

To obtain additional evidence of the coating of Al_2O_3 on Si_3N_4, the coated powders were studied by XPS. As shown in Table 2, the high Al concentration on the surface of Si_3N_4/Al_2O_3-B is an indication that Al_2O_3-B powder is more actively adsorbed. The low concentration of Al on Si_3N_4/Al_2O_3-A is in agreement with the ESA and pH_{iep} data. The presence of PAA was not particularly advantageous in promoting the attachment of Al_2O_3 to Si_3N_4. In fact, a decrease in the Al concentration indicates that PAA is not effective at this concentration (100 ppm). Uniformity of the Al_2O_3 coating was evaluated by the determination of Al concentration at various locations of the sample. The Si_3N_4/Al_2O_3-B system showed higher uniformity than the PAA containing system.

CONCLUSIONS

A technique for homogeneous distribution of Al_2O_3 as a sintering aid in the Si_3N_4 aqueous suspension was developed. The measurement of ESA was found to be instrumental in the characterization of Al_2O_3 coating on Si_3N_4 powder surface. An Al_2O_3 powder of extremely fine size (40 nm) was found to be highly effective in developing surface coatings on the Si_3N_4 powder. The 0.2 μm Al_2O_3 powder did not form as uniform a coating as that of 40 nm powder. The coating characterization by XPS confirmed the ESA data, in that the 40 nm particles produced a thicker and more uniform coating.

REFERENCES

1. M. Kulig, W. Oroschin and P. Greil, "Sol-Gel Coating of Si_3N_4 With Mg-Al Oxide Sintering Aid," J. Eur. Cer. Soc., 5, 209-217, 1989.
2. J. S. Kim, H. Schubert and G. Petzow, "Sintering of Si_3N_4 With Y_2O_3 and Al_2O_3 Added by Coprecipitation," J. Eur. Cer. Soc., 5, 311-319, 1989.
3. R. de Jong, R. A. McCauley, "Incorporation of Sintering Aids in Si_3N_4 By Means of Surfactants," in Processing of Ceramics, Vol. 1, Eds. G. de With, R. Terpstra and R. Metselaar, Elsevier Appl. Sci., 150-154, 1989.
4. S. G. Malghan, P. Pei and P. S. Wang, "Interface Chemistry of SiC Platelets During Alumina Coating" Accepted for Cer. Eng. and Sci. Proc., Paper 10-CP-91F, presented at Composites and Advanced Ceramics, Orlando, FL, February 1991.

ANALYSIS OF SURFACE CHEMISTRY OF SILICON NITRIDE AND CARBIDE POWDERS

S. G. Malghan, P. T. Pei and P. S. Wang
Ceramics Division, National Institute of Standards and Technology, Gaithersburg, MD

ABSTRACT

The surface chemistry of fine powders can be determined by several techniques, where each technique provides a different type of information. In the present study, two Si_3N_4 powders and a SiC powder were studied by surface sensitive x-ray photoelectron spectroscopy (XPS) and acoustophoretic mobility. The powders were exposed to oxidizing environments, similar to that existing in powder processing. The variation of oxide layer thickness and the pH at which the powder surface carries a net zero charge (pH_{iep}) were examined. The results indicated a slight increase of surface oxide thickness by exposing the powders to air, and the pH_{iep} also became more alkaline. After exposure to acidic or alkaline environments, all three powders showed a change in the surface oxide layer and pH_{iep}, the degree of variation was powder dependent.

INTRODUCTION

Surface Alteration of Si_3N_4 and SiC Powders

The Si_3N_4 and SiC powders are known to undergo surface oxidation when exposed to oxygen containing environments.[1] Since the oxygen content of powders affects composition and amount of glassy phase in the densified ceramics, an understanding of the oxidation behavior and control of the oxide content is necessary. Moreover, the surface oxidation affects interfacial chemistry of these fine powders which in turn is known to affect the powder dispersion behavior, rheology of suspensions, homogeneity of sintering aid distribution and consolidation of suspensions. In this study, an evaluation of surface characteristics of Si_3N_4 and SiC powders were examined by XPS and interface sensitive acoustophoresis techniques. The primary goal of this study was to develop powder surface chemistry data through surface alteration of fine powders by exposing them to different types of processing related environments.

38

EXPERIMENTAL

Three powders selected for this study were: A. Si_3N_4 LC-10 Starck*; B. Si_3N_4, Ube SNE-10; C. SiC, Starck. Size distribution data as determined by Sedigraph of A, B and C powders yield: $d_{50} = 0.76, 0.51$ and 0.47 μm, respectively. The powders were treated as follows:

I. As-received powders were sealed in argon ambient and stored for a year.

II. As-received powders were kept in closed plastic containers and stored for a year.

III. Powders in I were aged for 48 hrs., in a 10^{-3}N NaCl solution at pH 3.5 or 9.5 using 1N HNO_3 or NH_4OH.

The samples treated in this manner were studied by acoustophoresis and XPS. Acoustophoretic analysis of 0.5 volume % slurries were carried out using Matec 8050 analyzer. The XPS of dry powders was determined by a modified AEI ES-100 photoelectron spectrometer with a Mg Kα radiation. Samples for XPS were prepared by sprinkling on an adhesive tape.

RESULTS AND DISCUSSION

Surface Composition Determination by XPS

From the overall XPS scan of SiC and Si_3N_4 powders treated under above mentioned conditions, the peaks for O, C, N and Si were clearly observed. Figure 1 is an example of high resolution C 1s XPS spectrum from the surface of SiC sample treated at pH = 3.5. This signal was deconvoluted into three peaks at 282.9, 284.8 and 286.5 eV in binding energy for SiC, free carbon or hydrocarbon, and C-O, respectively. The O 1s electron binding energy was found to be 532.5 eV which is in good agreement with a reported value for SiO_2.[2] The Si 2p shows two peaks at 102.9 and 100.3 eV for oxide and carbide, respectively (Fig. 2). The surface composition in atomic % of all samples is presented in Table 1. A striking feature is high surface oxygen and carbon concentration on Si_3N_4-B in Argon and air-exposed samples. Although there is a binding energy difference of 2.6 eV in Si 2p of carbide and oxide, this difference for nitride and oxide was found to be less than 1.5 eV, which results in a poor resolution.[3] Therefore, use of Si 2p for separation of silicon nitride and oxide peaks is inappropriate. Since, Bremsstrahlung excited Si KLL Auger line has a separation of 3.2 eV and, it is more accurate for oxide layer thickness calculation.

---

* Certain trade names and company products are mentioned in the text or identified in illustrations in order to adequately specify the experimental procedure and equipment used. In no case does such identification imply recommendation or endorsement by National Institute of Standards and Technology, nor does it imply that the products are necessarily the best available for the purpose.

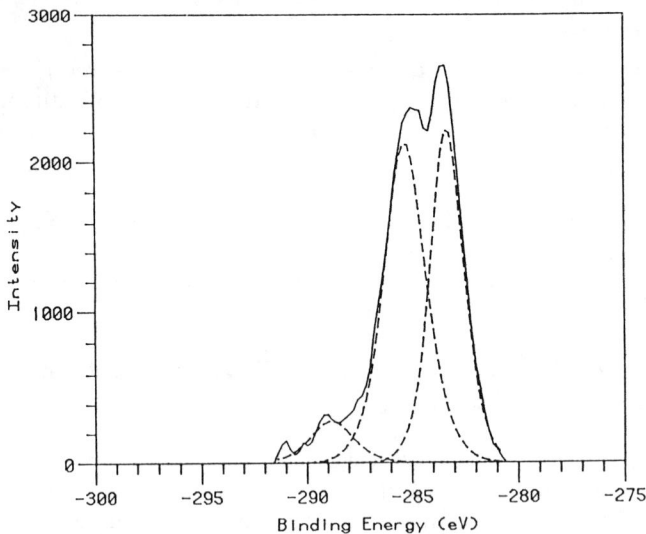

Figure 1 High resolution C 1s XPS spectrum of SiC treated at pH 3.5. Three peaks at 282.9, 284.8 and 286.5 eV in binding energy for SiC, free C or hydrocarbon and C-O are observed.

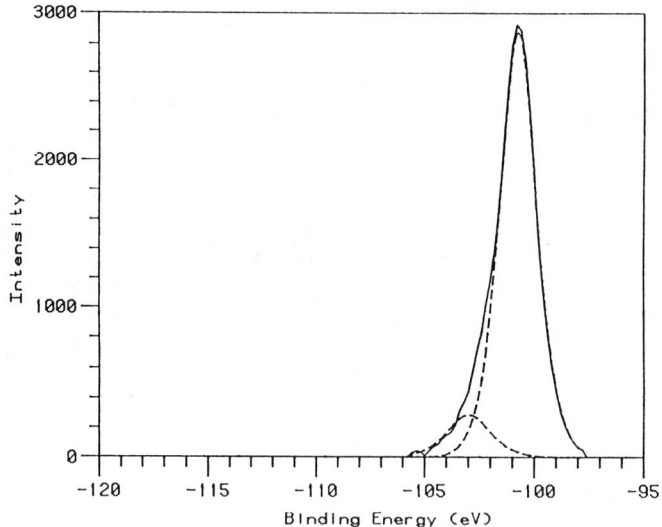

Figure 2 High resolution Si 2p XPS spectrum of SiC treated at pH 3.5. Peaks at 102.9 and 100.3 eV are due to oxide and carbide, respectively.

Table 1 Approximate Surface Composition (% Atomic) of Powders by XPS Due to Exposure to Different Environments.

Powder Exposed to	C			O	Si		N	O/Si
	C-O	CH	Si-C		Si-O, Si-N	SiC		
Si₃N₄-A								
Argon	—	11.6	—	16.5	30.3	—	41.4	0.54
Air	—	23.8	—	16.0	26.1	—	34.1	0.61
pH-3.5	—	32.3	—	21.9	22.6	—	23.1	0.97
pH-9.5	—	32.1	—	20.8	22.1	—	25.1	0.95
Si₃N₄-B								
Argon	—	29.7	—	23.3	22.5	—	24.5	1.03
Air	—	34.7	—	23.2	21.2	—	22.7	1.09
pH-3.5	—	33.0	—	16.3	22.2	—	28.5	0.73
pH-9.5	—	26.5	—	17.4	24.8	—	31.3	0.70
SiC								
Argon	—	27.2	27.6	14.9	3.4	25.6	1.3	0.51
Air	—	33.6	22.7	15.5	3.8	23.4	0.9	0.57
pH-3.5	3.5	26.9	26.1	15.3	2.6	25.7	—	0.55
pH-9.5	4.3	34.9	19.8	15.7	2.0	23.3	—	0.62

The surface oxide thickness of SiC or Si_3N_4 powders, t, was calculated from:[4]

$$t = -\lambda \ln[R/(R + R^*)]$$

where λ is the inelastic mean free path of Si KLL Auger electron, R is the Si KLL intensity ratio of carbide (or nitride) to oxide, and R^* is this ratio for pure samples. In these experiments, R^* is 1.39 for carbide and 1.22 for nitride and λ is 3.8 nm.[5,6] The surface compositions of approximately 10 nm deep were measured for each sample (Table 2). From this table, we can see that all three powders exhibit a slight increase in the thickness of Si-O layer by exposing to non-Argon environment. However, the major difference is in the Si-O layer thickness of Si_3N_4-B over that of Si_3N_4-A, which is an indication that powder B contains higher concentration of surface oxygen in as-received form. By treatment with acidic or alkaline pH, the oxygen content of Si_3N_4-B remains higher than that of Si_3N_4-A. At pH 9.5, powder A contains a slightly more oxygen than at pH 3.5. Powder B showed approximately the same oxide layer thickness at pH 9.5 and 3.5. The oxide layer on SiC powder was not affected at both pH.

Table 2 Thickness (nm) of Oxide Layer of Si_3N_4 and SiC Powders by XPS When Exposed to Different Environments.

Exposure to	Si_3N_4 -A	Si_3N_4 -B	SiC
Argon	0.44	0.89	0.50
Air	0.49	0.95	0.59
pH - 3.5	0.46	0.94	0.37
pH - 9.5	0.65	0.84	0.34

Interface Chemistry Characterization by Acoustophoresis

The influence of exposure to oxygen and aqueous environments on the acousto-phoretic mobility (electrokinetic sonic amplitude, ESA) and isoelectric point (pH_{iep}) are shown in Figures 3-5. Both Si_3N_4 powders showed a shift in pH_{iep} to a slightly higher pH due to exposure to oxygen-containing ambient (Fig. 3 and 4). Though the shifts are small, they are indicative of the changes in the surface characteristics resulting from the adsorption of cationic species.[7] The pH_{iep} and ESA of SiC powder over a wide pH range (Fig. 5) were unaffected.

When treated in acidic and alkaline pH solutions, all three powders behaved

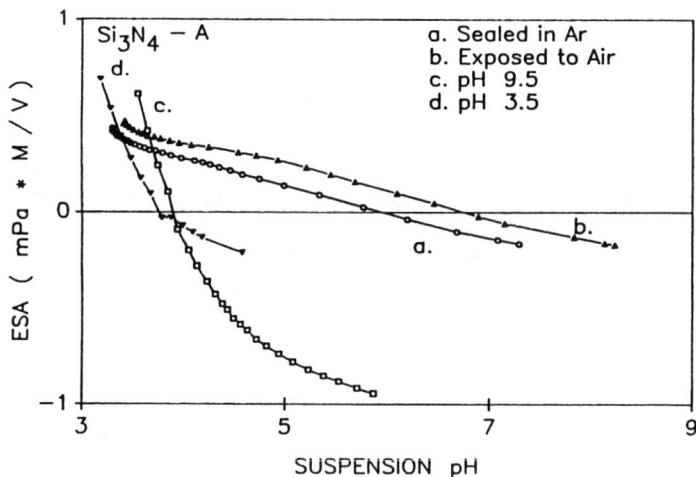

Figure 3 Acoustophoretic mobility of Si_3N_4-A by exposure to different oxidizing environments.

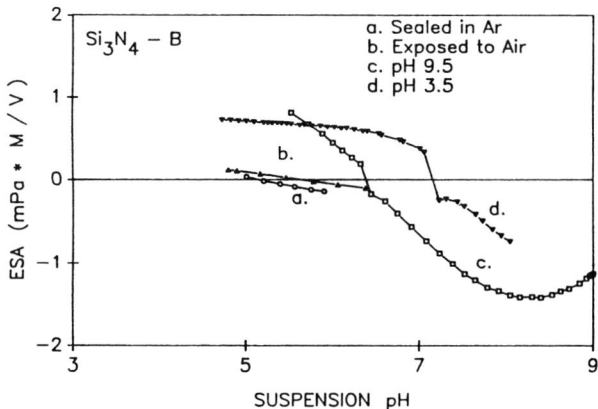

Figure 4 Acoustophoretic mobility of Si_3N_4-B by exposure to different oxidizing environments.

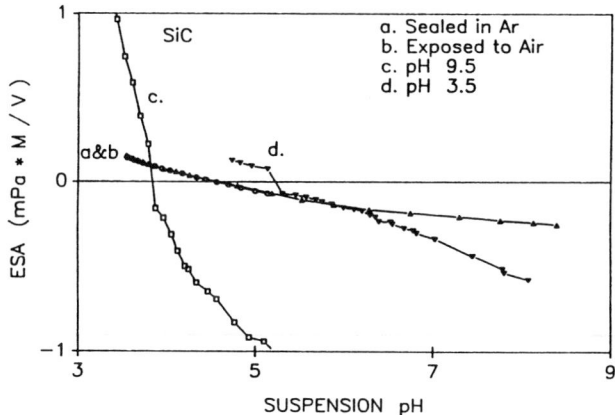

Figure 5 Acoustophoretic mobility of SiC by exposure to different oxidizing environments.

differently. In the case of Si_3N_4-A, the pH_{iep} shifted to a acidic pH, irrespective of the treatment pH. The shift in pH_{iep} was of the order of 2.5 pH units from that of Ar-sealed powder. On the contrary, the pH_{iep} of Si_3N_4-B shifted to an alkaline pH by treatment with pH 3.5 and 9.5 solutions. Further, the pH_{iep} shift of the powder treated at pH = 3.5 was larger than that of the powder treated at pH = 9.5. These data indicate that the surface of powder A interacts in a different manner in acidic and alkaline solutions compared to that of powder B. However, it is not clear if the observed difference is due to ionic species (Na^+, Cl^-) of supporting electrolyte in the treated solutions or due to the real differences resulting from ionic species of the powder surface. To eliminate the influence of ionic species of the supporting electrolyte, the Si_3N_4-B powder was treated in pH = 9.5 solution containing only NH_4OH. The pH_{iep} of the resulting powder was 6.4, which is nearly the same (6.3) as that of the powder treated in $10^{-3}N$ NaCl solution at pH 9.5. Therefore, the shift in pH_{iep} of Si_3N_4-A powder is closely related to the ions released from its surface.

The pH_{iep} of SiC powder treated in pH 3.5 and 9.5 solutions were 5.2 and 3.9, respectively. The treatment pH is expected to have an influence on surface dissolution which affects concentration and composition of the ionic species at the interface. Readsorption of the dissolved species, and exposure of the acidic surface could be responsible for the observed difference in the pH_{iep}.[7,8]

COMPARISON OF XPS AND ACOUSTOPHORESIS DATA

Though the two types of measurements reveal different information of the surface features, certain aspects are related. For example, the surface oxide thickness of all three powders is slightly larger by exposure to air. Similarly, the pH_{iep} of the Si_3N_4 powders are also slightly greater by exposing to air, indicating surface oxidation. In addition, O/Si ratio is different by exposing to either pH 3.5 or 9.5 for all three powders, when compared with air or Argon-exposed samples. Similarly, the pH_{iep} of these samples indicate that the degree of oxidation has changed by exposing to acidic or alkaline pH. Moreover, the pH_{iep} data demonstrate a varying degree of influence of cationic and anionic species resulting from surface dissolution.

CONCLUSIONS

Measurement of XPS and ESA of Si_3N_4 and SiC powders exposed to different oxidizing environments show that:

- By exposure to air, the surface oxide thickness increases very slightly; whereas, the pH_{iep} becomes more alkaline only for Si_3N_4 powders.

- By exposure to acidic and alkaline pH environments all three powders exhibited different oxide layer thicknesses and pH_{iep}. No generalized behavior was observed since different surface species come into play at different pH's.

ACKNOWLEDGMENTS

Authors thank Dr. D. Bartenfelder for carrying out some of the acoustophoresis measurements.

REFERENCES

1. P. K. Whitman and D. L. Feke, "Colloidal Characterization of Ultrafine SiC and Si_3N_4 Powders" Adv. Cer. Mtls., 1(4), 366-70, 1986.
2. T. N. Taylor, "The Surface Composition of Silicon Carbide Powders and Whiskers: An XPS Study", J. Mat. Res., 4(1), 189, 1989.
3. C. D. Wagner, "Practical Surface Analysis", Appendix 4, D. Briggs and M. P. Seah (eds.), John Wiley and Sons, New York, 1983.
4. P. S. Wang, S. M. Hsu and T. N. Wittberg, "Oxidation Kinetics of SiC Whiskers Studied by XPS", J. Mat. Sci., 26 1655, 1991.
5. C. J. Powell, "Energy and Material Dependence of the Inelastic Mean Free Path of Low Energy Electron in Solids", J. Vac. Sci. Technol., A3, 1338, 1985.
6. P. S. Wang, S. M. Hsu, S. G. Malghan, and T. N. Wittberg, "Surface Oxidation Kinetics of Si_3N_4-4% Y_2O_3 Powders Studied by Bremsstrahlung-Excited Auger Spectroscopy", J. Mat. Sci., in press, 1991.
7. M. A. Anderson and D. T. Malotky, "The Adsorption of Protolyzable Anions on Hydrous Oxides at the Isoelectric pH", J. Colloid Int. Sci., 72, 413-427, 1979.
8. S. G. Malghan, "Dispersion of Si_3N_4 Powders--Surface Chemical Interactions in Aqueous Media" Submitted to Colloids and Surfaces, Feb. 1991.

THE ROLE OF SURFACE TENSION IN THE FORMATION OF DONUT-SHAPED GRANULES DURING SPRAY-DRYING

K.J. Konsztowicz
National Research Council of Canada, Halifax, N.S. B3H 3Z1

G. Maksym, T. Maksym, H.W. King, and W.F. Caley
Technical University of Nova Scotia, Halifax, N.S. B3J 2X4

E. Vargha-Butler
Dalhousie University, Halifax, N.S. B3H 3J5

ABSTRACT

Highly concentrated suspensions of Al_2O_3 powder in aqueous solutions of poly-(vinyl) alcohol binder of molecular weight 25000 and varying content were prepared at pH 4. Viscosity and apparent surface tension of the slurries were measured at room temperature. Materials were spray-dried, and the morphology of the products was systematically analyzed by scanning electron microscopy. The explanation of the observed shape variability of spray-dried granules was based on combined models of liquid stream necking and falling drop formation. A springback effect of the elongated droplets due to excessive surface tension was found to be responsible for the "donut" shape formation of spray-dried granules of ceramic powders.

INTRODUCTION

The cause of formation of hollow granules during spray-drying remains uncertain. Internal voids of varying size, including "balloonlike" granules, are believed to appear during the spray-drying of low solid content slurries. These defects may also occur due to the presence of occluded air, and/or formation of an elastic film of low permeability at the droplet surface causing a temperature increase of the interior, with resulting internal evaporation of the moisture.[1,2] Another type of frequently observed hollow defect, the inward cavity, which at the extreme leads to "donut" shaped granules, seems to occur independently of the solid content of spray-dried slurries. Various explanations have been presented regarding the origin of the inward collapse of the granule surface, depending on the type of spray-dried material; however these are difficult to generalize.[1]

The purpose of the present communication is to introduce the apparent surface tension of a spray-dried slurry as a parameter determining the conditions for the formation of "donut" shape of spray-dried granules.

EXPERIMENTAL PROCEDURE

Materials

The details of the preparation of Al_2O_3 powder[*] used have been presented else-where.[3,4] Highly concentrated (53 vol% solids) colloidal aqueous suspensions of this powder, containing small amounts (0.3 %-1.27 wt %) of poly(vinyl) alcohol binder[†] with average molecular weight of 25000, were prepared at pH 4. After homogenization and defoaming, samples of each composition were given an identical ultrasonic agitation treatment[‡] (6 min, 300 W).

Methods

Viscosities of dense suspensions of submicron alumina in PVA binder aqueous solutions were measured at room temperature using a standard rotary viscometer.[§] The values of apparent surface tension of these suspensions were determined by the du Noüy ring technique[¶] at 25°C. Dense suspensions were spray-dried in a laboratory spray-dryer[**] and the morphologies of spray-dried products were systematically analyzed under the scanning electron microscope.[††]

RESULTS AND DISCUSSION

Fig. 1 shows the variations of the measured surface tension and viscosity of Al_2O_3 aqueous suspensions as functions of increasing content of PVA binder with the average weight 25000. It has been observed earlier[4] that when the viscosity is maintained at the level allowing for an easy flow of the slurry through the nozzle of the spray-dryer (<1000 mPa•s), and the apparent surface tension is close to its lowest value for a given MW (here, ~52 mN/m at 0.64 wt % of PVA, Fig. 1a), the compressed air can break the slurry into small regular volumes of mostly granular shape, as illustrated in Fig. 1b.

Only under ideal processing conditions can spray-dried droplets become close to spherical in shape and be fully dense. Under slightly different conditions (including variations of viscosity and surface tension of the slurry[4]), they can be still dense but elongated, remain spherical but with internal voids ("balloonlike"[1,2]), or be even

---

*   Reynolds, RCHP-DBM w/o MgO, d_{50} = 0.39 μm.
†   Airvol, Air Products and Chemicals Inc.
‡   Heat Systems, model W-385.
§   Brookfield, model LVTD.
¶   Krüss tensiometer, model K8.
**   Buchi, model 190.
††   Jeol, model JSM 820.

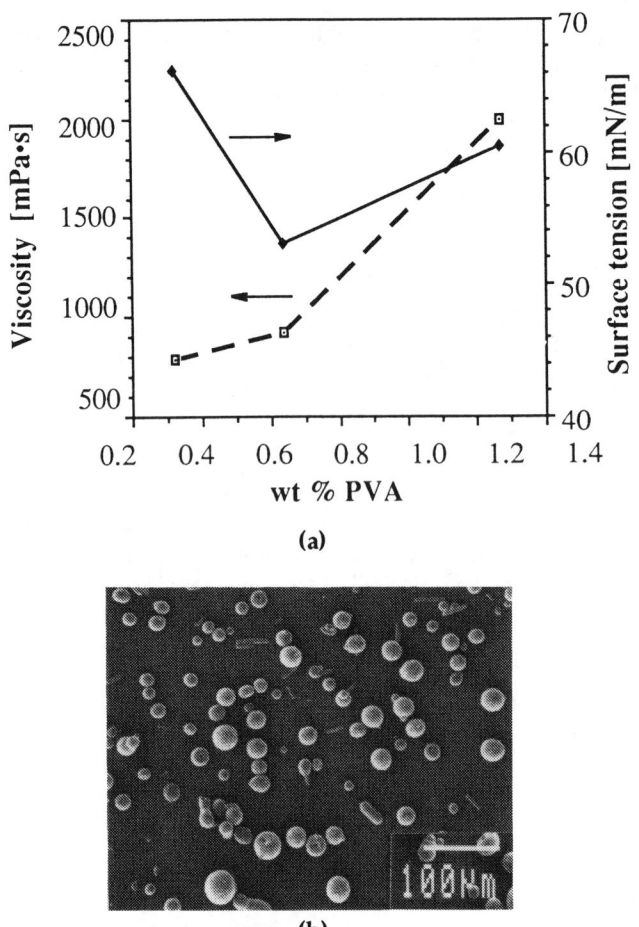

(a)

(b)

Figure 1 (a) Apparent surface tension and viscosity of dense slurry vs PVA binder content (MW 25000), (b) regular shape of granules spray-dried from the slurry containing 0.64 wt% of PVA (γ at minimum).

more deformed. In some cases the granules with a specific defect known as the "donut shape" may dominate. It is difficult to strictly classify various morphological defects of spray-dried droplets and relate them to particular processing conditions, since the final degree of surface deformation at solidification depends not only on the actual slurry characteristics, but also on many other processing parameters. Some of them are even impossible to determine, like the position of a given droplet relative to the drying hot air current at the moment of solidification. In fact, different

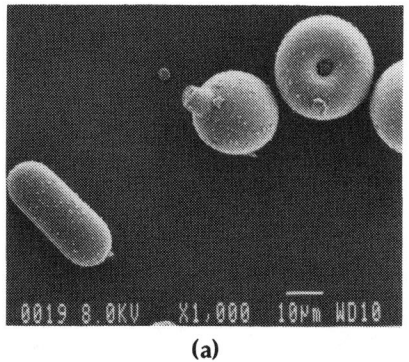

(a)

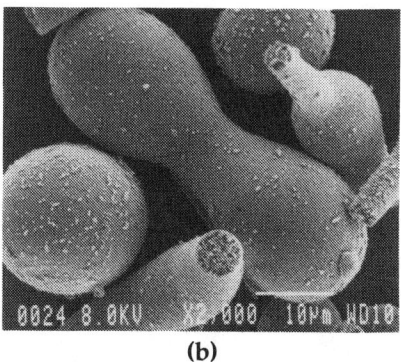

(b)

Figure 2 Different shapes observed within the same batch of spray-dried material: (a) elongated granule, granule with residual tail and "donut" shaped granule (b) necking of the elongated granule.

shapes were observed within the same batch of material spray-dried under identical conditions (Fig. 2).

Systematic morphological studies of spray-dried products indicate that their shape variability is strongly related to the surface tension of the starting slurry.[4] In the absence of gravity and other external stress fields, the droplets of any liquid should be spherical. External forces promoting deviations from spherical shape will increase the surface area of the drop, and consequently, the associated surface tension. Therefore, the shape of any drop of any liquid will always depend on the specific equilibrium conditions between the external stress fields introducing deformations, and the surface tension counteracting these changes.[5] Analytical solutions describing drop shape[5,6] consider either equilibrium conditions (sessile or pending drop),

49

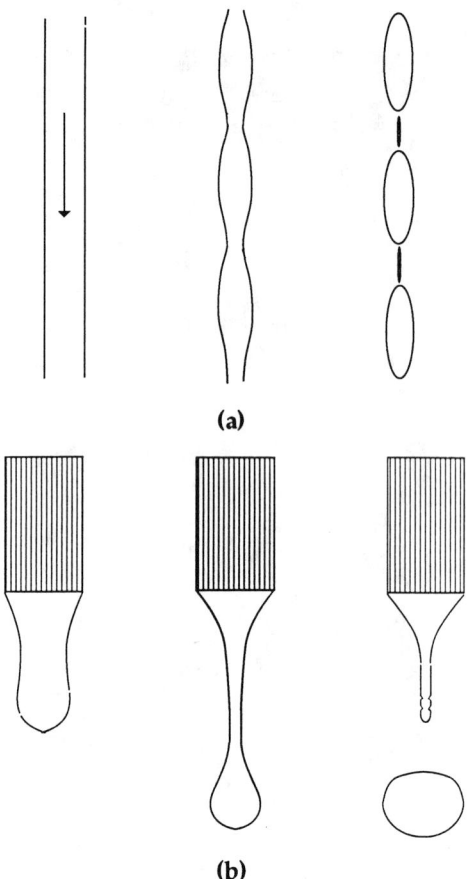

(a)

(b)

Figure 3 (a) Model of drop formation by necking and following disruption of a liquid stream (based on Fig. 1-7 from ref. 6); (b) stages of falling drop formation, extracted from series of high speed photographs (based on Fig. 1-14 from ref. 6).

or slowly varying conditions (falling drop), whereas in the case of spray-drying the conditions of drop formation are highly dynamic. The droplets being formed during spray-drying are subjected to the simultaneous action of multiple stress fields, the most important of which are gravity and the forced flow due to the action of compressed air. In this latter case, the formation of a multitude of liquid mini-streams can be expected, with all accompanying dynamic phenomena, including necking of these streams due to surface instability.[6] This leads directly to the disruption of a continuous liquid stream and to the formation of individual droplets of bimodal size distribution (Fig. 3a).

The "falling drop" formation model considers gravity as a major external stress field affecting its shape. The effects of gravity have to be taken into account particularly in the case of drop formation from highly concentrated ceramic suspensions. Gravity tends to lower the position of the mass center, which results in flattening of the bottom part of the drop (Fig. 3b). This effect increases the surface area, and consequently the surface tension. The equilibrium shape depends on the balance of these forces, as indicated by Tate' law in the form modified by Harkins and Brown:[6]

$$W' = 2\pi r \gamma f(r/a)$$

where: W' = actual drop weight
γ = surface tension of the liquid
r = radius of the tube from which the drop falls
a = capillary constant for a given liquid/tube configuration
(values of $f(r/a)$ are tabulated)

When the weight of the liquid prevails over the surface tension, the new drop is being formed, as shown schematically in Fig. 3b. The last case of this sequence indicates a very important change in drop shape: at the moment of detachment a strong effect of surface springback occurs, following the tendency to minimize the surface-to-volume ratio of gravitationally deformed droplet. This effect visibly flattens the top apex of a newly formed drop. On the basis of these observations it is reasonable to believe that the same (or even stronger) effect of springback of the drop surfaces should occur in the case of necking of the liquid stream under forced flow conditions during spray-drying (compare the shape of droplets in Fig. 2b).

It is expected that during spray-drying a multitude of slurry mini-streams are being formed at the nozzle orifice as a result of the action of compressed air. These streams are all subjected to a rapid necking due to stream surface instability (Fig. 3a); thus a significant degree of elongation of drops can be expected at the instant of their detachment (Fig. 4a). Some of these elongated droplets, before being dried, might undergo a gravitational mass displacement resulting in the formation of "droplets with tails" (see Fig. 2). In other cases of symmetrically elongated drops the excessive surface-to-volume ratio will promote a strong springback reaction by surface tension. This should lead to the inward collapse (of varying intensity) of drop apex surfaces. If at this instant the granule becomes dried by the hot air current, it will solidify into a "donut" shape (Fig. 4b).

CONCLUSIONS

Drop formation during spray-drying of ceramic suspensions is believed to occur due to surface instabilities and resulting processes of necking and disruption of liquid streams created in the nozzle.

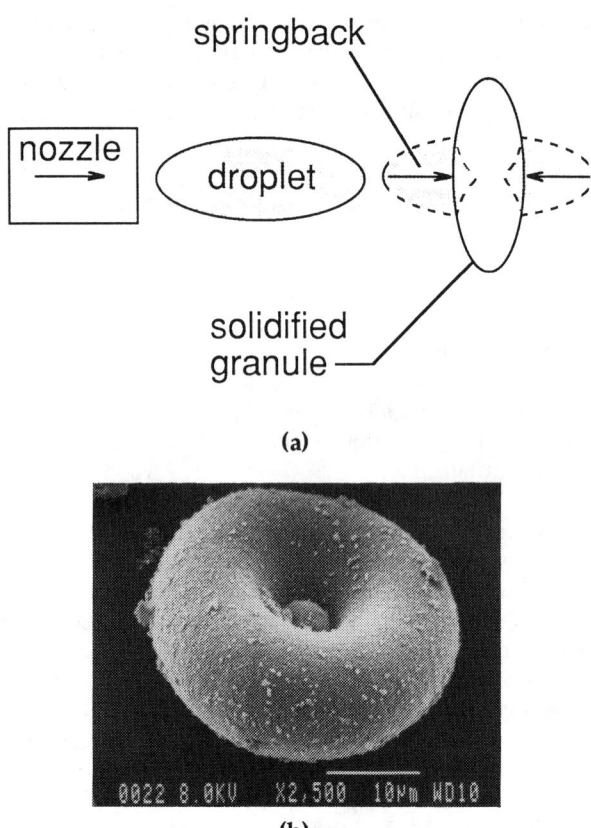

(a)

(b)

Figure 4 (a) Idealized model of springback effect due to droplet surface tension, leading to formation of the central inward cavity of the dried granule; (b) the real "donut" shaped granule.

Elongation of droplets formed from liquid streams, assisted by gravitational deformation, promotes a strong springback reaction due to surface tension tending to decrease the surface-to-volume ratio. This in turn leads to the inward collapse of opposite faces of the elongated drop, and the formation of a "donut" shaped dried granule.

Paper issued under NRCC No. 32975.

REFERENCES

1. S.J. Lukasiewicz, "Spray-Drying Ceramic Powders", J.Am.Ceram.Soc. 72 [4] 617-24 (1989).
2. K. Uematsu, J.Y. Kim, M. Miyashita, N. Uchida, K. Saito, "Direct Observations of Internal Structure in Spray-Dried Alumina Granules". J.Am.Ceram.Soc., 73 [8] 2555-57 (1990).
3. K.J. Konsztowicz, G. Maksym, H.W. King, "Spray-Drying of Dense Aqueous Suspensions of Alumina", pp. 1103-11 in "Ceramics Today--Tomorrow's Ceramics", Ed. P. Vincenzini, Elsevier Science Publishers, B.V., 1991.
4. K.J. Konsztowicz, E. Vargha-Butler, T. Maksym, H.W. King, "The Effect of Apparent Surface Tension on the Shape of the Products of Spray-Drying". To appear in Ceramics Transactions, 1991, Eds S. Hirano and G. Messing, Proc. of the 4th Int. Conf. on Ceram. Powd. Proc., Nagoya, Japan, March 1991.
5. P.C. Hiemenz, "Principles of Colloid and Surface Chemistry. Second Edition Revised and Expanded". Marcel Dekker Inc., New York and Basel, 1986.
6. A.W. Adamson, "Physical Chemistry of Surfaces" (IIIrd edition) J. Wiley and Sons, New York, 1976.

IN-SITU LIGHT SCATTERING STUDY OF AGGREGATION

Young Hoon Rim and James D. Cawley
Department of Materials Science and Engineering, The Ohio State University,
2041 College Rd., Columbus, OH 43210-1178

Rafat R. Ansari and William V. Meyer
NASA Lewis Research Center, 21000 Brookpark Rd., Cleveland, OH 44135

ABSTRACT

The results of both static and dynamic light scattering experiments on the aggregation of colloidal particles is reported. Commercial sols of silica and aluminum monohydroxide were employed. Aggregation was induced either by increasing salt concentration of a sol or mixing sols with oppositely charged particles, i.e. mutual flocculation or heterocoagulation. The agglomeration process was characterized by determination of the mean hydrodynamic radius and fractal dimension. The results of experiments on salt induced agglomeration of silica were consistent with prior studies. Mutual flocculation results yield a fractal dimension close to that predicted from the ballistic cluster cluster aggregation model.

INTRODUCTION

The process of aggregation or the flocculation of small particles to form larger clusters and the structures that result are important technologically and scientifically. A complete characterization of aggregation involves describing the kinetics of the process as well as the geometrical distribution of particles within individual aggregates. Dynamic, or quasielastic, light scattering (DLS) and static light scattering (SLS) provide complementary information towards this goal. DLS, which measures the time dependence of intensity fluctuations in the scattered light, has proven quite useful in study of the kinetics of aggregation and in the determination of particle size distributions.[1,2] Static light scattering, in which the angular dependence of time averaged intensity is measured, provides information on the internal structure of the aggregates. Both types of scattering experiments may be performed on the same sample and in-situ during aggregation.

Recently, the aggregation of colloidal gold and silica has been extensively studied and analyzed in the framework of fractal geometry.[3-12] In the context of colloidal aggregates, fractal geometry may be summarized with the following relations involving the fractal dimension, D:[13,14]

$$R_g \propto M^{1/D} \tag{1}$$

$$g(r) \propto \frac{1}{r^{(d-D)}} \qquad a \ll r \ll R_g \tag{2}$$

where R_g and M are the radius of gyration and mass of the aggregate respectively, $g(r)$ is the pair correlation function, a is the radius of the primary particles, and d is the spatial dimension ($d=3$ for all experiments discussed in this paper). The fractal approach to the description of aggregate structure has been recently fully discussed in the context of sol-gel processing of ceramics.[15]

The structures of experimental aggregates may be compared to those predicted from numerical models in order to infer the mechanism(s) of growth. The numerical models may be broken down into two principal classes: particle-cluster or cluster-cluster.[16,17] In the former, aggregate growth is considered to result only from the addition of singlets to an existing cluster, or aggregate, while the latter allows collisions between growing aggregates. Each of these broad classes may be further refined to account for Brownian motion (diffusion limited models), convection (ballistic models) or the presence of a significant barrier to aggregation (reaction limited models). Typically, the experimental fractal dimension is compared to the values from various numerical simulations and when a match is found it is inferred that the mechanism included in the model dominated under the given experimental conditions.

The kinetics of aggregation can be analyzed by following either the time dependence of R_g, i.e. the upper limit of Eqn-2, using SLS or through a determination of the hydrodynamic radius using DLS.

In DLS the initial decay rate of the fluctuations in the scattered light intensity may be determined from a cumulant analysis of the measured autocorrelation function. The ratio of this decay rate to the square of the wave vector may be used to obtain a measure of the translational diffusion coefficient which in turn may be analyzed using the Stoke's Einstein equation to yield a hydrodynamic radius, R_h. The wave vector, q, is defined by

$$q = (4\pi n/\lambda_0) \sin(\theta/2) \tag{3}$$

where n is the index of refraction, λ_0 is the wavelength of the light, and θ is the laboratory scattering angle. DLS studies have indicated that R_h follows an exponential time dependence under reaction limited conditions (slow aggregation)[9] and a power law under diffusion limited conditions (fast aggregation).[11]

Static light scattering may be analyzed to yield R_g by analyzing the transition between the power law regime and the Guinier regime. The scattered light intensity, $I(q)$, is directly proportional to the static scattering factor $S(q)$,[13,18] i.e.

$$\frac{I(q)}{I_0} = M^2 S(q) \tag{4}$$

Given an appropriate expression for $S(q)$, it is possible to analyze the measured intensity as a function of q, or equivalently θ, to determine R_g and D. One such expression which has been used in the analysis of colloidal silica aggregates[7,12] is

$$S(q) \propto \left[1 + \frac{2(qR_g)^2}{3D} \right]^{-(D/2)} \tag{5}$$

The function was selected based on its simplicity and the fact that it shows the correct limiting cases, which are:

$$I(q) \propto q^{-D} \qquad\qquad qR_g \gg 1 \tag{6}$$

and

$$I(q) \propto \left[1 - \frac{(qR_g)^2}{3} \cdots \right] \qquad qR_g \ll 1 \tag{7}$$

Since the R_h and R_g can be independently determined it is possible to evaluate the relationship between the two. Both experiments and simulations[12] suggest that the two are related by a simple proportionality and that the proportionality constant is near unity (although it has been pointed out that this analysis neglects the effect of polydispersity).

We used a slightly different approach. It is possible to relate $S(q)$ to a volume integral involving $g(r)$[13,14] and therefore an assumed form for $g(r)$ can be used to obtain the expression for $S(q)$. Since the integration volume is large compared to the size of the cluster it is necessary to multiply Eqn-2 by a cutoff function, $f(r)$, to account for finite size of the cluster. The form of this cutoff function must be such that $f(r) = 1$ for $r \ll R_g$ and $f(r) \to 0$ for $r \gg R_g$. Several functions have been investigated[12,13] including a step function, a Gaussian decay, $f(r) = \exp(-(r/R_g)^2)$, and an exponential, $f(r) = \exp(-r/R_g)$. We have employed the latter. Assuming $g(r) = \exp(-r/R_g)/r^{(d-D)}$

and performing the appropriate integration, with the assistance of an appropriate handbook,[20] yields

$$S(q) \propto \sin[(D-1)\tan^{-1}(qR_g)]\left[\frac{R_g^{(D-1)}}{q(1+(qR_g)^2)^{(D-1)/2}}\right] \quad (8)$$

Although this expression has a more complicated form than Eqn-5 substitution of the appropriate series obtains the same limiting forms.

The results of the SLS and DLS may be compared in two ways. Firstly, a relationship can be assumed between R_h and R_g may be assumed (e.g. $R_h \approx R_g$). This allows the data from DLS to be used to define R_g leaving only a single fitting parameter, D, for the SLS data. Alternatively, the data from DLS and SLS may be independently analyzed and the values of R_g and R_h compared.

EXPERIMENTS

Light scattering experiments were carried out on specimens in which aggregation was induced by one of two qualitatively different mechanisms: i) electrostatically stabilized single phase (silica) sols were rendered unstable through the addition of salt leading to aggregation as a consequence of van der Waals attraction or ii) two sols containing particles of opposite charge (negative silica and positive aluminum monohydroxide) were mixed leading to mutual flocculation, i.e. aggregation as a result of electrostatic attraction.

Materials

Scattering samples were made through dilution of commercial sols of either colloidal SiO_2 (Ludox-AM, DuPont Co., Wilmington DE) or AlO(OH) (Nyacol AL-20, Nyacol Products Inc., Ashland MA) using doubly distilled and deionized water (resistivity of ≈ 2.3 MΩ-cm). Ludox-AM is supplied as a sol of 30 wt.% (15.7 vol.%) silica. The particles are nearly spherical and closely sized. The surface of these particles have been modified by substitution of aluminum into the tetrahedral sites normally occupied by silicon. This produces a negative charge which is insensitive to pH. The suspension is supplied at a pH of 8.8, but was adjusted to 4 after dilution using nitric acid. Most experiments on salt induced aggregation were conducted at 0.5 wt.% which corresponds to a particle number density of 2.34×10^{15} cm^{-3}. The sol was diluted to 1.0% and an equal volume of a 2M NaCl solution was added yielding sample which was 0.5% SiO_2 and 1M NaCl. In some experiments the sample was agitated after the addition of the NaCl while in others extreme care was taken not to vibrate or shake the sample.

The Nyacol-AL20 is supplied as an electrostatically stabilized suspension (iep$\approx$8 and as-supplied pH $\approx$4) with 27.6 wt.% (11.2 vol.%) AlO(O)OH). In contrast to the

Ludox, AlO(OH) particles are high aspect ratio plates ($\approx$5 nm x $\approx$50 nm x $\approx$50 nm) and have a wider size distribution. For the mutual flocculation experiments, this suspension was diluted using a pH=4 nitric acid solution, and vigorously mixed with an equal volume of equally diluted Ludox.

To remove dust, sols were multiply filtered using either a LID/X filter syringe with a 0.2 μm membrane or a 0.2 μm Miller-FGS filter unit. The characterization of stable dilute sols included the determination of particle size distribution using DLS and confirmed using transmission electron microscopy.

All light scattering experiments were carried out using a commercial LLS system (Brookhaven Instrument Corporation) at the Laser Light Scattering Laboratory of the NASA Lewis Research Center. Our experiments employed a range of scattering angles from 20° to 150° for the SLS experiments. Most DLS data was collected at 90°. A vertically polarized 10 mW Ar laser (λ_0 = 5145Å) was used as the incident light source. In some experiments the incident beam intensity was reduced using neutral density filters. The scattered light was detected using a photomultiplier mounted on a precision goniometer. The reported SLS data represents the average of five 5-second runs at each scattering angle. The time necessary to complete an angular scan ($\approx$15 min) was short compared to the aggregation rate observed in the salt induced experiments and long relative to that observed in the mutual flocculation experiments (i.e. the aggregate structure is not expected to change on the time scale of the scan in either case).

In the DLS experiments a second cumulant analysis was used to determine the mean aggregate size as a function of time. The reported hydrodynamic radius must be regarded as an *apparent* hydrodynamic radius since the measurements were taken at 90° and therefore did not satisfy the criterion that $qR << 1$. It is known that failure to account for the q-dependence of the apparent value of R_h will result in an underestimate for a polydisperse system.[11,19]

RESULTS AND DISCUSSION

Salt Induced Aggregation of Ludox

Two types of experiments were performed. Our initial experiments were designed to grow large aggregates in order to minimize finite size effects. In these experiments, samples were prepared by allowing an agitated mixture of the diluted Ludox-AM and NaCl solution to rest undisturbed for a period on the order of a month. In all cases sedimentation was observed. The samples were resuspended using an ultrasonic probe and a specimen was extracted for SLS measurements. Power law behavior is observed over the entire range of scattering vectors and D was determined to be 2.21. This relatively high value for D is consistent with similar experiments on Ludox.[8,11]

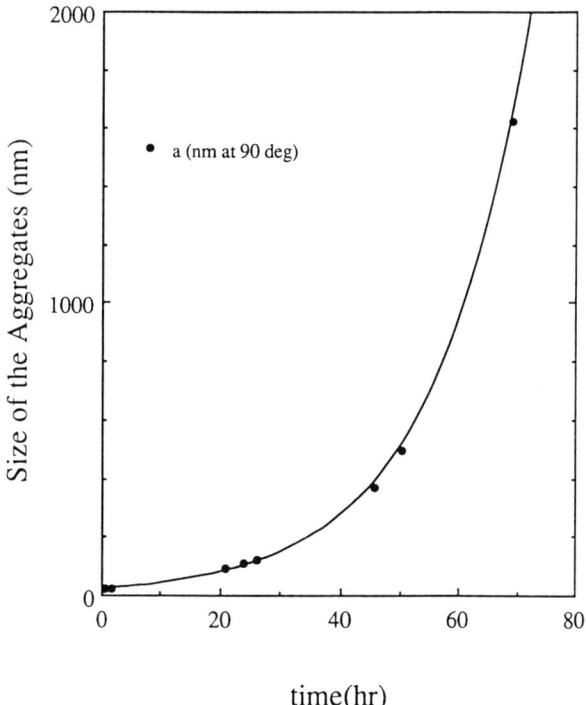

Figure 1 The value of the apparent hydrodynamic diameter determined from DLS on a salt induced aggregated silica 0.5 wt. % sol as a function of time. Zero time is defined by the addition of a salt solution to produce 1 M NaCl. The kinetics are best described by an exponential growth law.

The second set of experiments involved quiescent samples. In these experiments the 2 M NaCl solution was slowly poured into a cuvette containing the diluted Ludox. The cuvette was then placed directly into the light scattering sample holder and both SLS and DLS measurements were performed.

The kinetics of aggregation were found to follow an exponential growth law which is characteristic of reaction limited cluster aggregation.[11] The results from DLS experiments on a 0.5 Wt.% Ludox-AM sols with 1M NaCl is shown in Fig. 1. The apparent hydrodynamic diameters, a, presented in this figure were derived from a second cumulant analysis. The solid line represents a best fit to the displayed data which may be written as

$$a = 23.1 \exp(0.062 \bullet t) \tag{9}$$

where t is time expressed in hours and a is expressed in nm. The value of the constant in the argument of the exponent was not analyzed, but it is related to the concentration of particles and the magnitude of the residual repulsive interparticle forces.

It is surprising that exponential kinetics are observed since the experimental conditions (i.e. 1 M NaCl) were expected to yield fast diffusion limited aggregation and power law kinetics. This is particularly puzzling since the results of the SLS experiments indicate that the system was indeed undergoing diffusion limited aggregation.

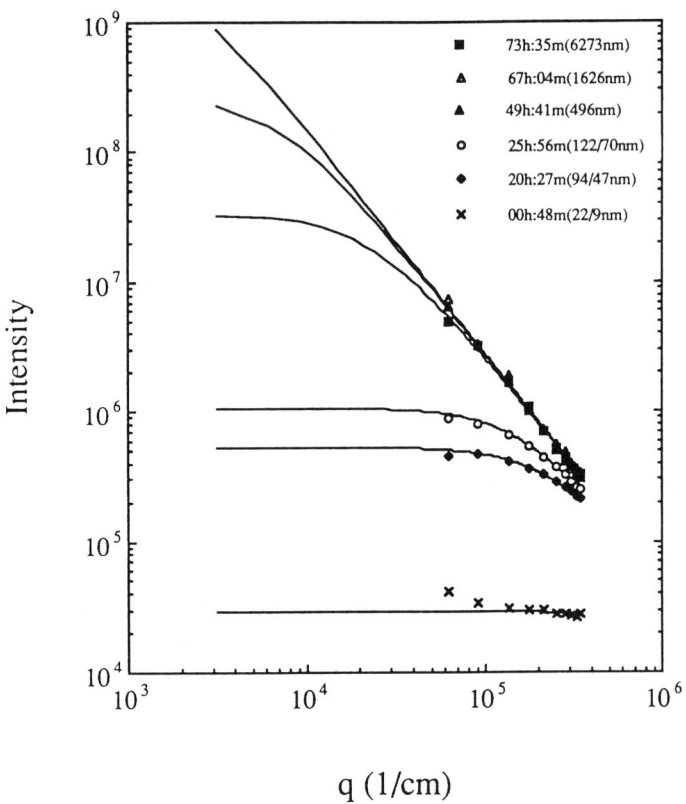

Figure 2 SLS profiles on the same sample as Fig. 1 as a function of time. The solid lines represent fits using Eqn-8 in the text for the static scattering factor, $S(q)$, with $D = 1.75$ and R_g assumed to be equal to one half of the hydrodynamic diameter determined for DLS. The hydrodynamic diameter for each fit is shown in the legend.

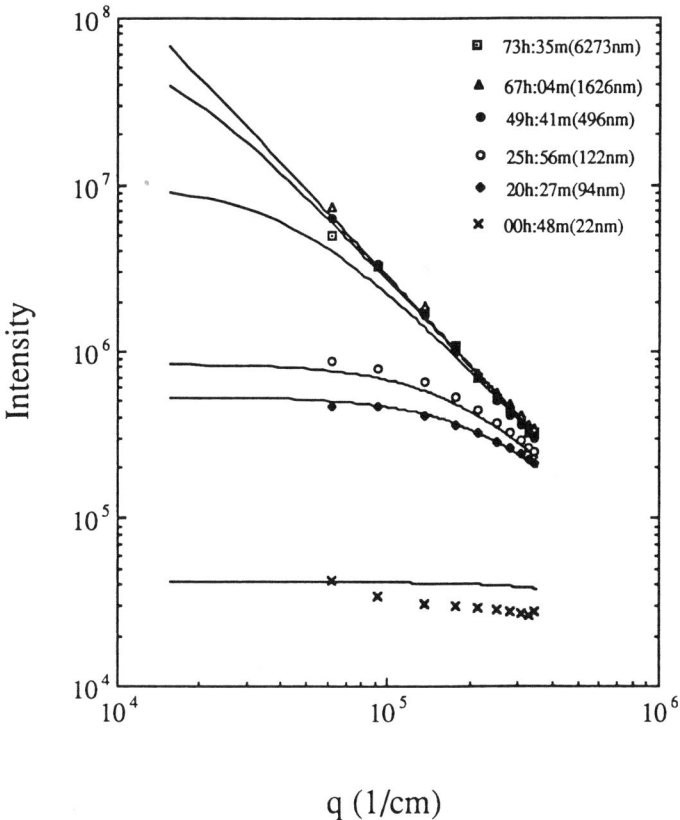

q (1/cm)

Figure 3 The same data as Fig. 2, but the solid lines are calculated using R_g as a fitting parameter in Eqn-8 (D still assumed to be 1.75). Significantly better fits are obtained particularly for times less than 49 hrs. R_g was assumed to be equal to the hydrodynamic diameter for the three uppermost curves and equal to the number to the right of the slash in the legend for the three lower curves.

The SLS results on the same sample are presented in Figs. 2 and 3. The same data points are plotted in each figure and the solid curves were all calculated using $S(q)$ as defined in Eqn-8 assuming $d=1.75$ which is very close to the value of 1.74 previously determined for fast diffusion limited aggregation of colloidal silica.[11] The difference between the two figures was the procedure used in evaluating R_g.

The curve in Fig. 2 were calculated assuming R_g was equal to the apparent R_h determined from DLS (one half of the hydrodynamic diameter given in parentheses in the legend). In order to calculate intensity it is necessary to determine the

Table 1 R_h and R_g Determined from SLS and DLS, Respectively.[a]

Time (hrs.)	R_h (nm)	R_g (nm)	qR_g	R_h/R_g
0.80	11	9	0.2	1.2
20.45	47	47	1.1	1.0
25.93	61	70	1.6	0.9
49.68	248	496	11.5	0.5
67.07	813	1626	37.8	0.5
73.58	3137	6273	145.9	0.5

[a]Values determined on a 0.5 wt. % sol of Ludox-AM in a 1 M NaCl solution as a function of time. The value of R_g was determined by fitting the data in Fig. 3 using Eqn-8. The value of R_h was determined from DLS at 90° using an Ar Laser yielding $q = 1/43$ (nm^{-1}).

magnitude of the proportionality constant in Eqn-8. This was empirically done for one set of data and the same value was used for all subsequent calculations. The calculated curves are in good, though not perfect, agreement with the experimental data both in terms of the systematic increase in intensity and the shift of break in slope to lower values of q. This indicates that the assumption of a simple proportionality between the characteristic aggregate sizes measured in DLS and SLS is a good approximation.

Somewhat better fits to the SLS data can be obtained if the value of R_g is used as a free fitting parameter. This was done in generating the solid curves presented in Fig. 3. A summary of the values of R_g determined from SLS and R_h determined from DLS as a function of the dimensionless parameter qR_g presented in Table 1 shows that when qR_g is on the order of 1 or smaller the ratio R_h/R_g is nearly unity and that when $qR_g \gg 1$ this ratio is ≈ 0.5 (it should be noted that the fit to the SLS becomes insensitive to R_g as aggregation proceeds and therefore small differences from 0.5 cannot be resolved). These results are consistent with the fact that the effective diffusion coefficient observed in DLS increases as qR_g exceeds 1 as a result of the increasing importance of rotational diffusion.[18,19] The magnitude of the effect, i.e. a factor of 2, is in very good agreement with the predictions that have been made for diffusion limited aggregation over the range of qR_g used in our DLS experiments.[18,19]

Mutual Flocculation of Ludox-Nyacol Mixtures

Mutual flocculation was observed to be very rapid. Even at particle concentrations of 0.1 wt. % the process was complete before the sample could be inserted into the light scattering system (1-2 min.). As a result of the rapid growth of large aggregates power law behavior was observed throughout the entire range of q investigated.

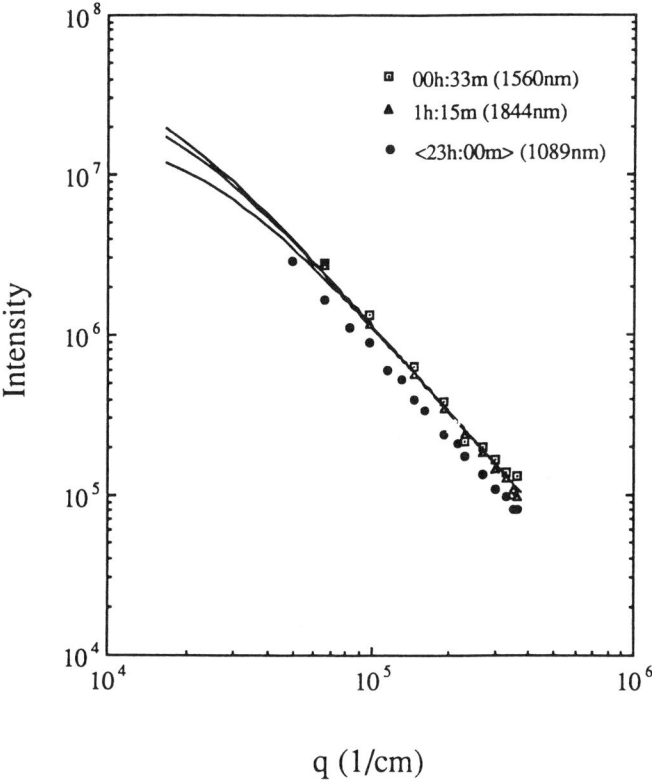

Figure 4 SLS profiles determined from a mixture of Ludox and Nyacol sols of 0.1 wt. % as a function of time. R_h values determined from DLS are given in the legend. Power law behavior is observed and the intensity decreases as a function of time which is attributed to large aggregates settling out of the path of the primary beam.

The results of a typical experiment using 0.1 wt. % suspensions are shown in Fig. 4. The best fits using Eqn-8 and the $2R_h$ from DLS yield $D = 1.84$ which is significantly higher than that observed for the salt induced silica aggregates. The value of D observed for mutual flocculation of 0.5 wt. % suspensions was significantly higher, ≈ 2.03. The only change in the SLS with time was a slow uniform decrease in the intensity of the signal, $\approx 40\%$ after 23 hrs. DLS indicates that the aggregate size first increases and then decreases. Both of these observations are consistent with settling under gravity. The net effect being the reduction of the number of scattering centers in the path of the beam as the larger clusters settle down.

The high fractal dimensions observed in the mutual flocculation experiments may be qualitatively rationalized in terms of electrostatic interactions. The fact that the aggregation kinetics are very rapid indicates that the electrostatic interaction is relatively long range. Thus it is expected that particle motion will not be dominated by Brownian motion, rather particles of opposite charge will tend to follow ballistic trajectories towards each other. Such a process is consistent with a higher fractal dimension.[16,17] In addition, it is plausible (though unproven) that electrostatic interaction between neighboring branches within an aggregate will increase as the aggregate grows and lead to restructuring. This would also be expected to drive D to higher values.

CONCLUSIONS

1) The assumption of an exponential cutoff on $g(r)$ yields a form of $S(q)$ which is consistent with the SLS results on both salt induced single phase aggregates and aggregates produced by mutual flocculation.

2) Silica aggregates grown in a 1 M NaCl solution yield $D = 1.75$. Aggregates allowed to settle and then resuspended exhibit $D = 2.21$. Both observations are consistent with prior work. The scaling relationship between R_g (determined by SLS) and R_h (determined from DLS) is also consistent with prior calculations and experiments.

3) Mixtures of silica and aluminum monohydroxide aggregate very rapidly and exhibit higher fractal dimensions suggesting ballistic aggregation and possibly restructuring.

ACKNOWLEDGMENTS

RA and WVM would like to acknowledge support from the NASA microgravity Science Application Division, Code SN, of NASA. YHR and JDC would like to acknowledge support through NASA grant #NAG3-755 and a seed grant from the Ohio State University Center for Materials Research.

REFERENCES

1. B. J. Berne and R. Pecora, Dynamic Light Scattering, Wiley, New York, 1976.
2. B. E. Dahneke, Measurement of Suspended Particles by Quasi-Elastic light Scattering, Wiley, New York, 1983, Ch.1.
3. D. A. Weitz and M. Oliveria, "Fractal Structures Formed by Kinetic Aggregation of Aqueous Gold Colloids", Phys. Rev. Lett. 52, [16] 1433-36 (1984).
4. D. A. Weitz, J. S. Huang, M. Y. Lin, and J. Sung, "Limits of Fractal Dimension for Irreversible Kinetic Aggregation of Gold Colloids," Phys. Rev. Lett, 54, [13] 1416-19 (1985).
5. D. W. Schaefer, J. E. Martin, P. Wiltzius, and D. S. Cannell, "Fractal Geometry of

Colloidal Aggregates", Phys. Rev. Lett. **52**, [26] 2371-74 (1984).

6. D.W. Shaefer and K. Keefer, "Fractal Aspects of Ceramic Synthesis," p. 277-88 in <u>Better Ceramics through Chemistry II</u>, Edited by C.J. Brinker, D.E. Clark, and D.R. Ulrich, Mat. Res. Soc. Symp. Proc. **73** (1986).

7. D. S. Cannell and C. Aubert, "Aggregation of Colloidal Silica," p. 187-197 in <u>On Growth and Form: Fractal and Non-Fractal Patterns in Physics</u>, Edited by H. Eugene Stanley and N. Ostrowsky, Martinus Nijhoff Publishers, Boston MA, 1986.

8. C. Aubert and D. S. Cannell, "Restructuring of Colloidal Silica Aggregates," Phys. Rev. Lett. **56**, [7] 738-41 (1986).

9. J. E. Martin, "Slow Aggregation of Colloidal Silica," Phys. Rev. A **36**, [7] 3415-26 (1987).

10. H. M. Lindsay, M. Y. Lin, D. A. Weitz, P. Sheng, Z. Chen, R. Klein, and P. Meakin, "Properties of Fractal Colloid Aggregates," Faraday Discuss. Chem. Soc. **83**, 153-65 (1987).

11. J. E. Martin, J. P. Wilcoxon, D. Schaefer, and J. Odinek, "Fast Aggregation of Colloidal Silica," Phys. Rev. A **41**, [8] 4379-91 (1990).

12. P. Wiltzius, "Hydrodynamic Behavior of Fractal Aggregates," Phys. Rev. Lett. **58**, [7] 710-15 (1987). Z. Y. Chen, P. Meakin, J. M. Deutch, "Comment on 'Hydrodynamic Behavior of Fractal Aggregates'," Phys. Rev. Lett. **59**, [18] 2121 (1987). P. N. Pusey, J. G. Rarity, R. Klein, and D. A. Weitz "Comment on 'Hydrodynamic Behavior of Fractal Aggregates'," Phys. Rev. Lett. **59**, [18] 2122 (1987). P. Wiltzius and W. van Saarloos, Reply, Phys. Rev. Lett. **59**, [18] 2123 (1987).

13. J. E. Martin, "Static and Dynamic Scattering From Fractals," Phys. Rev. A **31** [2] 1180-82 (1985).

14. J. E. Martin and A. J. Hurd, "Scattering From Fractals," J. Appl. Cryst. **10** 61-78 (1987).

15. C. J. Brinker and G. W. Scherer, <u>Sol-Gel Science: The Physics and Chemistry of Sol-Gel Processing</u>, Academic Press, San Diego, CA, 1990.

16. P. Meakin, "Models for Colloidal Aggregation," Ann. Rev. Phys. Chem. **39**, 237-67 (1988).

17. R. Jullien and R. Botet, <u>Aggregation and Fractal Aggregates</u>, World Scientific, Singapore, 1987, p. 64, p.77.

18. D. A. Weitz and M. Y. Lin, "Laser Light Scattering as a Probe of Fractal Aggregates," p. 173-184 in <u>NASA Laser Light Scattering Advanced Technology Workshop-1988</u>, NASA CP 10033, Edited by W.V. Meyer.

19. H. M. Lindsay, R. Klein, D. A. Weitz, M. Y. Lin, and P. Meakin, "Effect of Rotational Diffusion on Quasielastic Light Scattering From Fractal Colloids," Phys. Rev. A **38** [5] 2614-26 (1988).

20. I. S. Gradshteyn and I. M. Ryzhik, <u>Table of Integrals, Series, and Products</u>, Academic Press, New York, 1980, p. 490.

RHEOLOGICAL DETECTION OF AGGLOMERATES IN CONCENTRATED ALUMINA SLURRIES

B.M. Moudgil and M.E. Springgate
Department of Materials Science and Engineering, University of Florida,
Gainesville, FL 32611

ABSTRACT

Presence of agglomerates in concentrated slurries is detected by an increase in the viscosity. In this study, viscosity measurements were used to assess the state of dispersion of 41 vol% alumina slurries. Flocs of known characteristics were employed to determine the limit of detection of flocs. It was found that depending on size, binding strength, and amount of flocs, measurable changes in viscosity may or may not occur. For example, 1.0 vol% of 0.15 mm diameter flocs with a binding strength of 11.0 dyne/cm^2 in a 41 vol% slurry did not change the viscosity. It is shown that conventional viscosity techniques may not detect the presence of soft agglomerates, thus limiting the reliability of these methods for quality control of concentrated slurries.

INTRODUCTION

The production of monolithic and composite ceramic bodies through the suspension processing (i.e. slip casting) of highly concentrated slurries is of major importance in the ceramic industry. Implementation of ceramics in critical applications produced in this manner has been limited due to the inability to produce large quantities of defect free ceramic bodies.[1]

Quality control of the casting slips is generally accomplished through rheological characterization of the slurries. The general rule being, the lower the viscosity of the slurry, the better the dispersion of the particles. This condition is expected to result in optimal particle packing and high densities and homogenous microstructures of the sintered bodies. However, in practice, slight variations in the dispersion of slurries caused by changes in powder characteristics, slurry pH and ionic strength, or changes in the dispersant dosage result in sintered bodies with non optimal properties.[2]

In this study, ability of viscosity measurements to detect changes in the dispersion of concentrated ceramic slurries was examined. Flocs (particle networks) were added to otherwise dispersed slurries and their presence was monitored by viscosity measurements.

EXPERIMENTAL PROCEDURE

Commercially available alumina powder (Alcoa, A-16sg) having an average particle size (d_{50}) of 0.5 microns and a surface area of 8 m^2/g was used. Surface charge characterization was conducted using both microelectrophoresis (Pen Kem, Lazer Zee Model 501) and electrokinetic sonic amplitude (ESA) analysis (Matec, Model 8000). Zeta potential values using microelectrophoresis were obtained for samples of 500 ppm solids, which were centrifuged from 41 vol% slurries so as to maintain a constant ionic strength at any given pH. ESA values were obtained from 41 vol% slurries at various pH values.

Soft agglomerates (flocs) of the alumina powder were produced in a flocculator which consisted of a standard reaction vessel (7.6 cm inner diameter, 14 cm tall) fitted with 0.85 cm wide baffles. Mixing was achieved with a six-bladed stainless steel turbine impeller attached to a stir controller (HST-10, G.K.G. Heller Corp.) for monitoring agitation speed and torque input into the reactor.[3] The binding strength of the flocs was determined from the torque input of the agitator using the following equations[4]

$$B.S. = \frac{\rho_w}{2}\left[\frac{\varepsilon^* d_f}{\rho_w}\right]^{2/3} \tag{1}$$

where: $B.S.$ = floc binding strength
ρ_w = density of the suspending medium
ε^* = energy input into the system
d_f = average floc diameter

$$\varepsilon^* = \frac{\tau g_c \omega}{V} \tag{2}$$

where: ε^* = energy input into the system
g_c = acceleration due to gravity
τ = torque input into the system
ω = angular velocity
V = volume of suspension

The floc size was determined by measuring the diameter of a statistically representative number of flocs using an optical microscope.

Table 1 Effect of Shear Speed on Floc Diameter and Binding Strength.

Shear Speed (rpm)	Floc Diameter (mm)	Binding Strength (dyne/cm^2)
900	0.30	6.5
1500	0.15	11.0

Slurries with controlled amounts of flocs were produced by preparing a dispersed slurry (at optimal pH conditions). Separately, a known quantity of alumina was flocculated. The flocs were washed to remove any residual polymer and then added to the dispersed slurry taking care to maintain a total solids loading of 41 vol%.

Rheological characterization was carried out using a rotating spindle viscometer (Brookfield, LVTD) over a range of shear rates from 2.04 to 20.10 sec^{-1}. A linear increase in viscosity, at a shear rate of 20.1 sec^{-1}, was observed (slope: 39.651 cps/solids wt%) in the solids range of 37 to 42 wt%. The reproducibility of viscosity measurements is within ± 10 cps.

RESULTS AND DISCUSSION

Surface charge characterization by both microelectrophoresis and ESA showed a maximum occurring at pH 4 (zeta potential=60 mV, ESA=10 mPa*M/V). Therefore, all subsequent slurries were prepared at pH 4.

Previous studies have shown the effects of pH on the characteristics of flocs.[5] To prevent any discrepancies, all flocs were formed and characterized at pH 4. Table 1 outlines the effect of agitation speed upon floc size and floc binding strength. From the results one can see that the floc size decreases and the binding strength increases as the rate of shear to which the flocs are subjected is increased. These results are in agreement with those of previous investigators.[4,6,7]

The viscosity results presented in Figures 1 and 2 are characteristic of electrostatically stabilized concentrated particulate slurries.[8-10] The non-Newtonian (pseudo-plastic) behavior is characteristic of concentrated suspensions where the rheological properties are influenced by the interactions of the flow fields which surround the individual particles.[11,12]

It is also observed that as the floc size decreases (increasing floc binding strength) at a constant amount (0.15 mm flocs and 0.30 mm flocs at 5.0 vol%) the ability of viscosity measurements to detect the presence of the flocs decreases. This is demonstrated by a change in viscosity of 125 cps in the presence of 0.15 mm diameter flocs

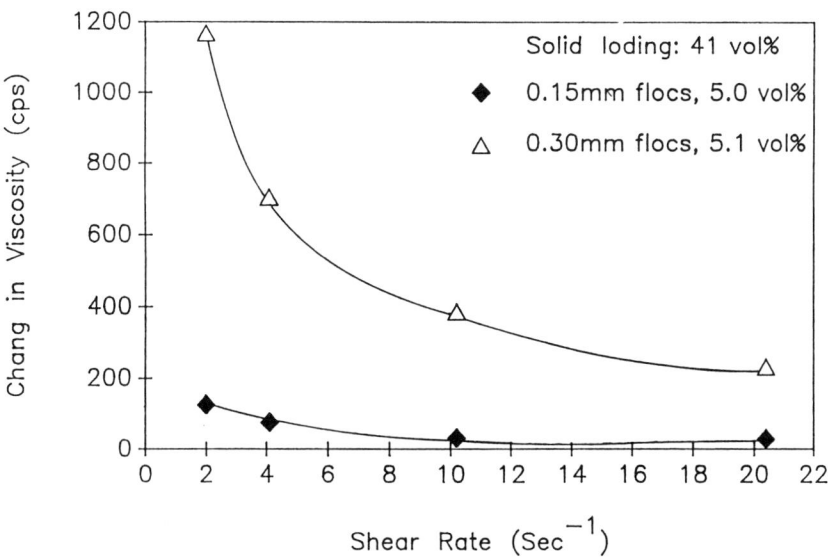

Figure 1 Viscosity behavior of 41 vol. % alumina slurry with 0.15mm diameter flocs.

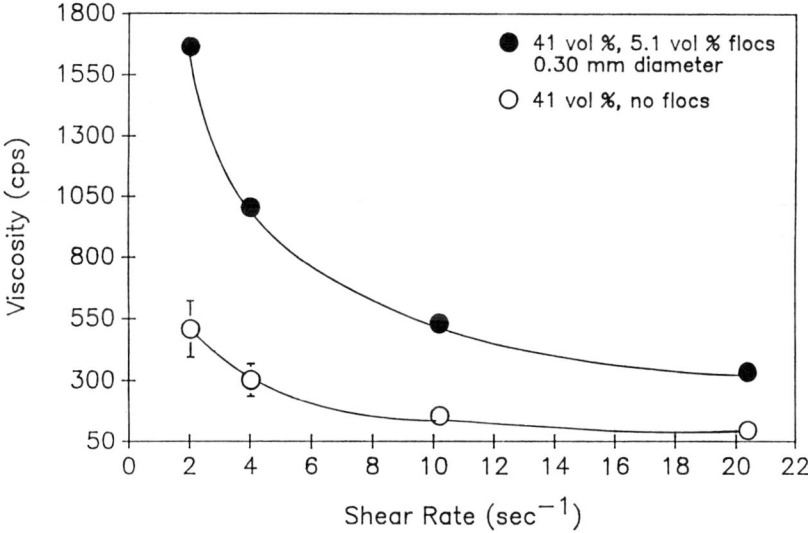

Figure 2 Viscosity behavior of 41 vol. % alumina slurry with 0.30mm diameter flocs.

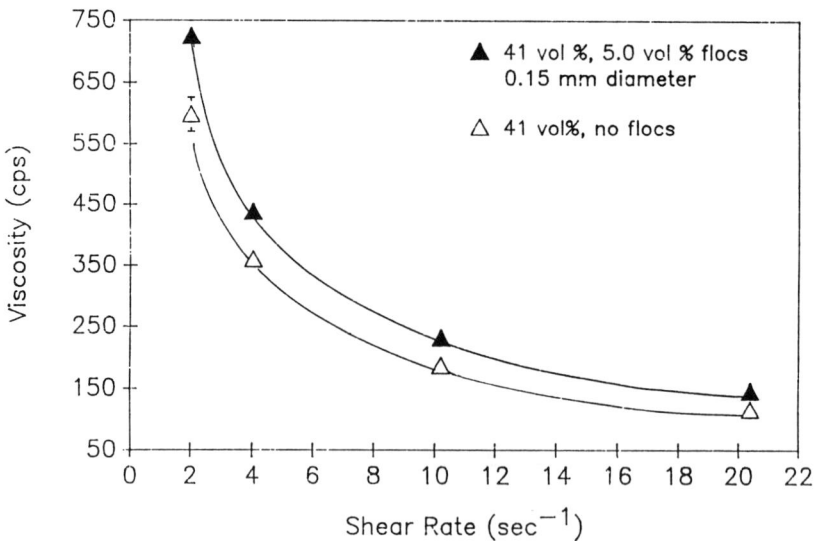

Figure 3 Change in viscosity of the slurry with flocs as a function of shear rate.

as opposed to 1150 cps for the 0.30 mm diameter flocs at a shear rate of 2.04 sec^{-1}. It is also noted that the ability to detect the presence of the flocs is enhanced at lower shear rates (see Figure 3). Thus, for detection of flocs in a concentrated suspension, the use of a lower shear rate would be fortuitous.

In order to establish a limit of detectability of flocs, subsequent experiments were carried out with varying amounts of 0.15 mm and 0.30 mm diameter flocs. It is seen from the results presented in Table 2 that the ability to detect the presence of flocs, even under low shear rates, is lost as the amount of flocs decreases. Flocs of a smaller size and higher binding strength have a lower limit of detectability as opposed to those of larger size and smaller binding strength. For example, flocs of 0.15 mm diameter do not increase the viscosity at 1.0 vol%, while flocs of 0.30 mm diameter do not go undetected until their quantity has been decreased to 0.2 vol%.

The increase in viscosity of the slurries in the presence of the flocs is attributed to entrapment of liquid within the flocs which increases the effective solids loading of the slurry. Experiments were conducted to determine the amount of liquid entrapped inside the flocs. For these experiments a representative number of flocs were separated from the slurry. The excess water was removed and the flocs were weighed while still wet. They were subsequently dried at 120°C and reweighed to determine the amount of liquid entrapped within each floc. It was found that flocs of 0.30 mm diameter entrapped 0.5 g H_2O/g floc and flocs of 0.15 mm diameter

Table 2 Viscosity of Slurries With and Without Flocs.[a]

Slurry	Viscosity (cps)
No Flocs	550
Flocs 0.15mm Diameter, 1.0 Vol. %	550
No Flocs	1050
Flocs 0.30mm Diameter, 0.2 Vol. %	1050

[a]Total Slurry = 41 Vol. % Solids; Shear Rate = 2.04 Sec. $^{-1}$

entrapped 0.3 g H_2O/g floc. The effective solids loading of the slurries with 0.2 vol% of 0.30 mm flocs and 1.0 vol% of 0.15 mm flocs are 41.1 vol% and 41.2 vol% respectively. This increase in solids loading is within experimental error range, which previous experiments had shown to be 20 cps, and is not sufficient to result in a measurable change in viscosity of the 41 vol% slurry.

CONCLUSIONS

Floc size and floc binding strength were controlled by the speed of agitation to which the flocs were subjected. It was shown that as the rate of agitation was increased, the floc size decreased, but the binding strength of the resultant flocs increased. Viscosity measurements carried out using 41 vol% slurries with varying amounts of 0.15 mm or 0.30 mm diameter flocs showed that as the amount of flocs of a given size and binding strength decreased, the ability of viscosity measurements to detect their presence was lost. It was also observed that at a lower rate of shear, detection of the flocs was enhanced. These results suggest that for quality control purposes, lower shear rates should be employed for better slurry characterization. The limitations of viscosity measurements in detecting the presence of soft agglomerates (flocs) and the need for developing other techniques to characterize the state of dispersion of concentrated slurries are clearly established.

ACKNOWLEDGMENTS

Financial support of this work by the National Science Foundation (grant no. MSS 8821815) is acknowledged.

REFERENCES

1. Dagani, K., "Ceramic Composites Emerging as Advanced Structural Materials," Chem. & Eng. News, (February 1, 1988) pp. 7-12.

2. Lange, F.F., "Powder Processing Science and Technology for Increased Reliability," J. Am. Ceram. Soc., 72, [1], 1989, pp. 3-15.
3. Springgate, M.E., M.S. Thesis, University of Florida, 1990.
4. Moudgil, B. M., and Vasudevan, T.V.," Evaluation of Floc Properties for Dewatering Fine Particle Suspensions," Minerals and Metallurgical Processing, 8, 1989, pp. 142-145.
5. Moudgil, B.M., Springgate, M.E., and Vasudevan, T.V.," Characterization of Flocs for Solid/Liquid Separation Processes," in Solid/Liquid Separation: Waste Management and Productivity Enhancement, Ed. A.S. Muralidhara, Batelle Press, Columbus, OH, 1989, pp. 245-253.
6. Tomi, D.T. and Bagster, D.F., "The Behavior of Aggregates in Stirred Vessels," Trans. I. Chem. E., 56, 1978, pp. 1-18.
7. Tambo, N. and Hozumi, H., "Physical Characterization of Flocs-II. Strength of Flocs," Water Research, 13, 1979, pp. 421-427.
8. Sacks, M.D., "Properties of Silicon Suspensions and Cast Bodies," J. Am. Ceram. Soc., 63, [12]. 1984, pp. 1510-1515.
9. Sacks, M.D., Lee, H.W., and Rojas, O.E., "Pressureless Sintering of SiC Whisker-Reinforced Composites," Ceram. Eng. Sci. Proc., 9, [7-8]. 1988, pp.741-754.
10. Cesarano, J. and Aksay, I.A., "Processing of Highly Concentrated Aqueous α-Alumina Suspensions Stabilized with Polyelectrolytes," J. Am. Ceram. Soc., 71, [12], 1988, pp.1026-67.
11. Mewis, J. and Spaull, A.J.B., "Rheology of Concentrated Dispersions," Adv. Colloid and Interface Sci., 6 1976, pp.173-200.
12. Tadros, Th. F., "Control of the Properties of Suspensions," Colloids and Surfaces, 18, 1986, pp. 137-173.

THE EFFECT OF DISPERSANT CONCENTRATION ON THE CAST CAKE STRUCTURE AND RHEOLOGY

I. Tsao and R. A. Haber
Department of Ceramic Science and Engineering, Rutgers University,
Brett and Bowser Roads, Piscataway, NJ 08855-0909

ABSTRACT

For an aqueous alumina slip (76 wt% or 42 vol%), the cake rheology was characterized in situ using Rheometric mechanical spectrometer. Results showed that partially flocculated and flocculated slips, containing 0.25 wt% and 0.28 wt% tetrasodium pyrophosphate (TSPP) based on solids, had higher viscosities but faster casting rates in comparison to deflocculated slips, containing 0.42 wt% TSPP. After casting, if the drying time is less than 30 minutes, flocculated slips resulted in higher elastic moduli and yield stresses compared to deflocculated slips. However, at a 60 minute drying time, the cake resulting from flocculated slips showed lower elastic moduli and yield stresses.

INTRODUCTION

Ceramists have studied the casting behavior and suspension rheology of aqueous alumina slips in efforts to achieve better solids dispersions and improved green densities. One can refer to Phelps et al.,[1] Smith and Haber[2] and Sacks et al.[3] for detailed information. However, in order to increase casting rate, partially flocculated slips are commonly employed as shown by Funk[4] as well as Kelly and Brodie.[4] This is because the flocs provide larger pores or larger channels for faster dewatering. Partially flocculated slips have also been shown to exhibit plasticity that is conducive to better mold release and cast trim. This phenomenon was attributed to higher moisture retention found in the flocculated system. In the casting process, demolding is a critical step. Poor demolding behavior can result from sticking and ultimately affect the structure of the cake leading to cracking. To successfully remove the part from mold, the rheological behavior of the consolidated body, i.e still containing moisture, plays an important role. These rheological properties can be correlated with the cake microstructure for optimal demolding.

Recently, Lange and Velamakanni et al.[5,6] studied the influences of interparticle forces on particle consolidation. As a slurry was flocculated by adjusting the pH of the suspension, they found that the effect of electrostatic repulsion was reduced due to the solvation of counterions on the particles and a coagulated microstructure is formed. This coagulation resulted in a weaker network structure than for the flocculated slip. In a subsequent investigation,[7] the cake rheology was characterized using an Instron type mechanical testing machine. A constant step strain of 2% was applied for 0.2 second to the pressure consolidated alumina particles. The stress responses were recorded to show different rheological behavior as a function of pH value. They showed that the ratio of the saturation stress (after relaxation) to the maximum stress (prior relaxation) was nearly zero at pH range of 3 to 4 (i.e. dispersed state) in contrast to a value as high as 0.7 at pH of 7 (i.e. flocculated state). This suggests that a flocculated cake could have more cracking than a dispersed system after pressure release. In addition to the experimental results, one of the most important implication of their work was the first exposure of cake rheology in the field of slip casting. Although the characterization was not conducted in situ, i.e. microstructure might be slightly destroyed during removing sample, the results indicate that the rheological characterization of cast cakes can be useful to correlate with their microstructures.

In this study, deflocculated, partially flocculated and flocculated alumina slips were selected to study the effect of drying time on demolding process. The rheological behavior of cast cake was carefully examined to resolve the microstructural difference between deflocculated and partially flocculated systems. In contrast to using an Instron type mechanical testing machine, the Rheometric mechanical spectrometer was used for the characterization of cake rheology. Its extremely sensitive transducer can detect the shear stress and shear strain relationship at a shear rate as low as $10^{-6} s^{-1}$. More importantly, all tests were conducted in situ without having to disturb the cake which could alter its microstructure.

EXPERIMENTAL PROCEDURE

Standard alumina slips, i.e. 76 wt% (or 42 vol%), were prepared with different dispersant concentration in double distilled water. The alumina powder used was A16SG.[*] Tetrasodium pyrophosphate[†] was dissolved in water to be used as dispersant. The ratio was fixed at 12 g TSPP/100 g water. The deflocculation curve was first obtained using Brookfield viscometer at fixed speed of 20 rpm. Based on the deflocculation curve (plotted in suspension viscosity vs dispersant concentration), slips A, B and C with different contents of TSPP were selected, i.e. 0.25, 0.28 and 0.42 wt%. Slip A had highest viscosity and was defined as the most flocculated slip, but it could still be poured into the mold. Slip B had higher viscosity than that of

---

* Alcoa Co., Pittsburgh, PA.
† Fisher, Inc., Fair Lawn, NJ.

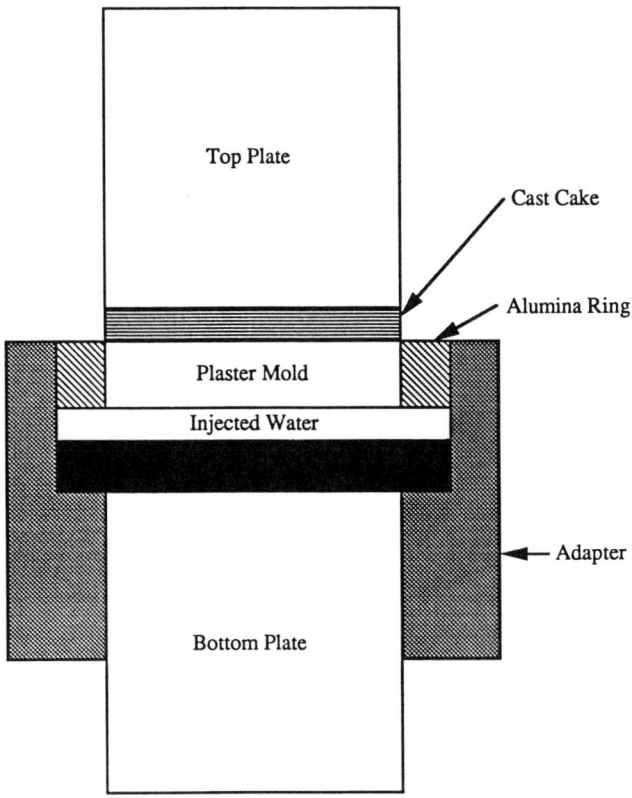

Figure 1 The schematic drawing of a pair of modified parallel plates.

deflocculated slip and was defined as partially flocculated slip. Finally, slip C was determined as deflocculated slip because it had the lowest suspension viscosity. After selection, all fresh slips were ball milled for 1 hour to further breakdown large agglomerates.

A Rheometric mechanical spectrometer was used to study the cake rheology in situ. A pair of parallel plates was used and modified as measurement tools. As seen in figure 1, the bottom plate was driven by the motor and connected with an adapter on which a 80 consistency plaster mold within a ring was held. Alumina slips were cast into an additional ring on the top of the plaster. The casting time varied from 30 to 90 seconds and the cake thickness was precisely controlled at 1.5 mm by adjusting the gap between two parallel plates. The plaster was then saturated with moisture by injecting water from the bottom of the plaster cell using a syringe. This

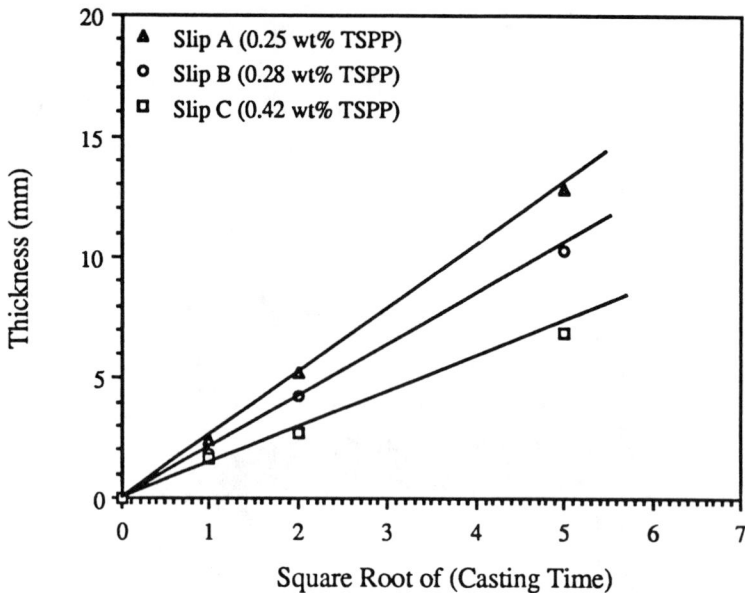

Figure 2 Cake thickness vs square root of casting time for slips A, B and C, T = 24°C.

step eliminated the remaining drawing pressure from the mold. Prior to conducting the measurement, different drying times were held at 2, 5, 10, 30, 60, 130 and 200 minutes, respectively, for slips A, B and C. Then, a constant shear rate of 10^{-4} s^{-1} was immediately applied for 5 minutes to study the cake structure in situ.

RESULTS AND DISCUSSION

The suspension viscosities of slips A, B and C were measured as 2200, 1200 and 500 mPa•s, respectively. Slip A was much thicker than slips B and C, however it still could be cast. Afterwards, the casting rates were determined in terms of cake thickness as a function of casting time for all three slips. Figure 2 shows that flocculated and partially flocculated slip result in higher casting rate, i.e. slip A was greater than B which was greater than C. This is because flocculation of slip actually increases the porosity and pore size of the cake, while the surface area remains the same. Therefore, according to Funk,[4] water can be more easily drained out for slips A and B, while the moisture distribution should be more uniform across the cake. More importantly, by using a mercury porosimeter, the average green bulk density of cake B was 2.24 g/cm^3 which was very similar to 2.30 g/cm^3 of cake C. This indicates that partially flocculated slip can result in faster casting rate than deflocculated slip, but with similar green bulk density. In addition, a flocculated (or partially flocculated) structure can prevent the blockage of the porous plaster surface from dispersed alumina particles. This blockage phenomenon may lead to mold blinding which can decrease the rate of casting and result in mold sticking.

To study the cake rheology, the influence of shear rate (i.e. at 10^{-4} s^{-1}) was monitored as a function of drying time. As a result, for slip A, figure 3 shows the shear stress as a function of shear strain and drying time. At 2 and 5 minutes, there is no apparent yielding phenomenon and the shear stress gradually increases with increasing strain. However, a linear range was observed for the remaining three drying times, indicating a pure elastic behavior region. A yield point (i.e. proportion limit) can be determined from the stress-strain curve, followed by a plastic deformation behavior due to the breakage of internal structure. Here, the yield stress is defined as an offset yield stress and the offset is specified to be 0.02%. The elastic modulus is the slope over the linear region of each curve. As a result, the yield stress and elastic modulus can be determined and usually increase with increasing drying time because of cast cake becoming more and more rigid. However, at 60 minutes, it shows much lower stress level compared to that of 30 minute curve. A short linear region is observed and quickly followed by a nearly steady state stress level instead of a plastic flow (i.e. gradually increasing stress). This phenomenon suggests that

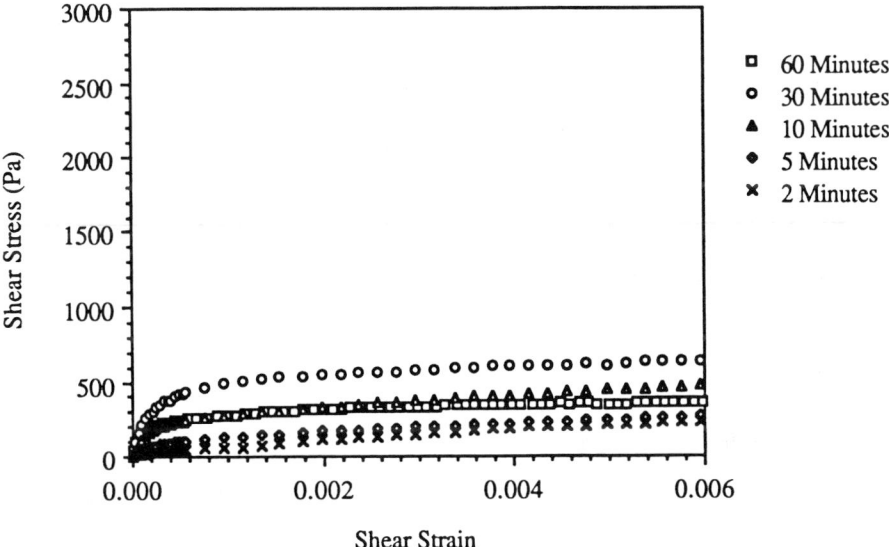

Figure 3 Shear stress vs shear strain for 1 minute cast cake of slip A, at different drying times.

the cake might be detached from the mold. This is supported by visual observation which shows that the cake can be easily demolded after 60 minutes drying with no or very little sticking. The same phenomenon was also found for slip B. The only difference is that the stresses level of slip B are higher than those of slip A for the same drying time. This can be explained due to better packing efficiency which leads to less moisture content in the cake at each stage.

For slip C (i.e. deflocculated slip), a different rheological behavior as a function of drying time is observed, compared to partially flocculated slips (i.e. slips A and B). In figure 4, except at 2 and 5 minutes, each curve exhibits a clear yield point and followed by plastic flow. However, the yield stress and elastic modulus increase with increasing drying time up to 200 minutes instead of being reduced in the cases of partially flocculated slips. These high stresses and moduli can be attributed to the best packing state of all three slips, where cast cake retains least moisture content. According to figure 4, the continuous increased yield stress is an indication that the cake is still firmly adhering to the mold. This sticking phenomenon was also verified by visual examination. As seen in the figure 5, the yield stresses resulting from different slips are also plotted against the moisture content (i.e. corresponding to different drying time) within the cake. At low moisture contents, the yield stress drop for slips A and B can be concluded as a signal of cake detachment at the mold

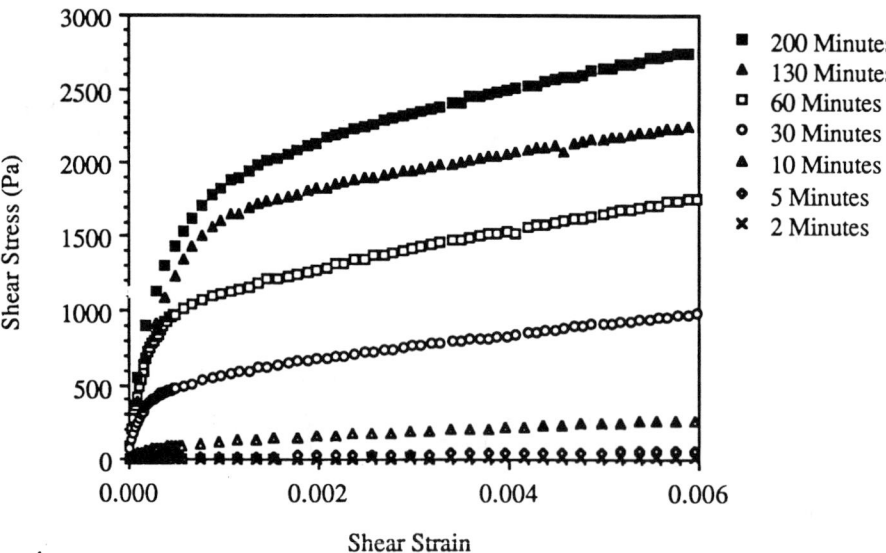

Figure 4 Shear stress vs shear strain for 1 minute cast cake of slip C, at different drying times.

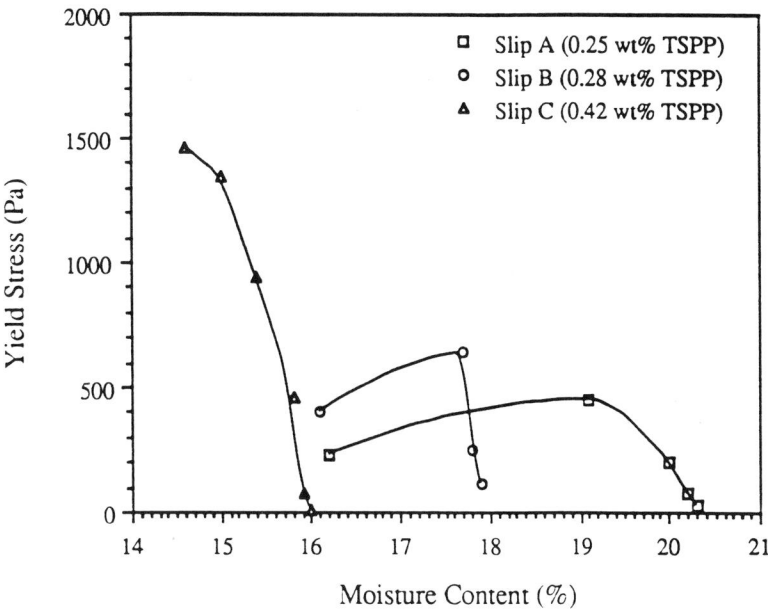

Figure 5 Yield stress vs moisture content for the cakes resulting from slips A, B and C.

interface. During test, the shear motion of these cakes should be a combination of partial deformation as well as finite slippage at the mold interface.

CONCLUSIONS

From above discussion, the rheological behavior of cast cake is very important and can be correlated with its microstructure during casting and demolding. Of importance, in this study, the cake rheology was characterized in situ using a Rheometric mechanical spectrometer. Results show that partially flocculated slurries, i.e. with 0.25 wt% or 0.28 wt% (based on totals solid) TSPP, have higher viscosities but faster casting rates in comparison to deflocculated slip, i.e. with 0.42 wt% TSPP. However, if the drying time is less than 30 minutes, flocculated slips resulted in higher yield stresses and elastic modulus compared to deflocculated slip. This suggests that the cake from partially flocculated system could be demolded and subject to a greater strain without affecting the structure. However, at 60 minutes, the cake resulting from flocculated slips shows lower yield stress and elastic modulus. This can be attributed to the evidence of cake detachment at the mold interface.

ACKNOWLEDGMENTS

The authors would like to acknowledge Mr. D. Italiano and M. Lalwani for their laboratory assistance and would also like to thank the Casting Technology Program of Center for Ceramic Research at Rutgers University for financial support.

REFERENCES

1. G. W. Phelps, S. G. Maguire, J. Kelly and R. K. Wood, "Rheological and Rheometry of Clay Water Systems," published by Cyprus Industrial Minerals Co., Sanderville, GA, 1983.
2. P. Smith and R. A. Haber, "Characteristics of Large Extensions in the Size Distribution for Alumina Slips," *Ceramic Engineering & Science*, **12** [1-2], 93-96 (1991).
3. M. D. Sacks, C. S. Khadilkar, G. W. Scheiffele, A. V. Shenoy, J. H. Dow and R. S. Sheu, "Dispersion and Rheology in Ceramic Processing," in Advances in Ceramics, Vol. 21: Ceramic Powder Science, Ed., G. L. Messing et al., Am. Ceram. Soc. Inc., Westerville, Ohio, 1986.
4. J. E. Funk, "Slip Casting and Casters," in Advances in Ceramics Vol. 9: Forming of Ceramic, Ed., J. A. Mangel et al., Am. Ceram. Soc. Inc., Columbus, Ohio, 1984.
5. F. F. Lange, B. V. Velamakanni, J. C. Chang and D. S. Pearson, "Colloidal Powder Processing for Structural Reliability: Role of Interparticle Potential on Particle Consolidation," in Proceeding of the 11th Riso International Symposium on Metallurgy and Material Science: Structural Ceramics-Processing, Microstructure and Properties, Ed. J. J. Bentzen et al., Denmark, 1990.
6. B. V. Velamakanni, J. C. Chang, F. F. Lange and D. S. Pearson, "New Method for Efficient Colloidal Particle Packing via Modulation of Repulsive Lubricating Hydration Forces," *Langmuir*, **6** [7], 1323-1325 (1990).
7. B. V. Velamakanni, F. F. Lange and D. S. Pearson, "Influence of Interparticle Forces on the Rheology of Pressure Consolidated Particulate Bodies," presented in American Ceramic Society meeting, Anaheim, CA, November 1989.

DILATANT TRANSITIONS OF ALPHA ALUMINA SLIPS

David A. Barclay
Vista Chemical Company, 12024 Vista Parke Drive, Austin, TX 78726-4050

ABSTRACT

Highly loaded slips of alpha alumina exhibit either dilatant or shear thickening behavior. Rheologic measurements are an effective means of investigating this transition. Techniques for studying the behavior are shown.

INTRODUCTION

Flow curves and oscillatory measurements on highly concentrated alumina slips reveal the changes in dispersion structure caused by dispersants and powder properties. A knowledge of the structure of the dispersion can indicate ways to increase the solids loading. Rheologic measurements can be used to confirm that the proper structure has been created in the dispersion prior to casting.

BACKGROUND

Previous workers[1-3] showed the importance of creating an alpha alumina dispersion that is completely deagglomerated and highly concentrated. Elimination of ordering and clustering of the alumina particles is obtained by using high volume percent concentrations of alumina with the correct dispersants. This prevents aggregation or flocculation that occurs in dilute concentrations, reducing long range ordering in the green body.

The importance of pseudoplastic behavior in the slip has been shown.[4,5] This prevents particle settling. It gives a predictable viscosity behavior to the slip for control of casting. Organic dispersants impart a controllable pseudoplastic behavior to the dispersion.

Other workers[6,7] showed the importance of flow curves for assessing the state of aggregation in an alumina slip. Rheological measurements can be used to study dispersion structure.[8] They were shown to be effective in determining the state of interaction between particles in a colloidal dispersion.

Alumina slips with high volume fraction exhibit a complex rheology. With increasing alumina volume fraction, a transition from shear thinning or pseudoplastic fluid behavior to either a dilatant fluid or a shear thickening fluid occurs.[5] This has been referred to as shear hardening.[6] The slip flows at low shear stress and viscosity dramatically increases at a critical stress.[5]

The viscosity of unagglomerated particle dispersions increases rapidly with volume fraction; several equations are available to describe this.[5] These are not successful for dispersions that form structures. Okamoto[9] showed the power law index of alumina dispersed with polyacrylic acids increased greatly near the maximum solids loading. It changed from an index for pseudoplastic flow to one indicating dilatancy.

EXPERIMENTAL

Two alpha alumina powders were used in this study. The first had 0.5 micron average particle size with a surface area of 7 m^2/g. The second powder had 0.4 micron average particle size with 12 m^2/g surface area. The surface areas were determined by the BET method; particle size by Sedigraph.

The technique for making dispersions was kept simple and avoided too many variables. Highly concentrated dispersions described by others were not attempted.[10] The powder and water with dispersant were blended with a spatula until a uniform dispersion was obtained.

The primary dispersant was the sodium salt of polymethacrylic acid (PMAA) with a molecular weigh of 5,000. Dispersions made with PMAA were kept near pH 9. The amounts added were determined from the work by Cesarano, et al.,[10] and are the optimum level found by them to completely coat the particles. For the 7 m^2/g alumina 0.002 g PMAA/g Al_2O_3 were used, for the 12 m^2/g 0.0035 g PMAA/g Al_2O_3 were used. Dilute nitric acid was used to make dispersions with a pH near 2.

Rheology of the dispersions was studied on a controlled stress rheometer. A parallel plate measuring system with a 500 micron gap was used. To prevent water evaporation from the sample edges, a cover was used over the measuring head which kept the space around the sample water saturated. All of the measurements were at 20°C.

RESULTS AND DISCUSSION

Figure 1 shows representative flow curves obtained during the study. All were run as non-equilibrium flow curves. For most of these systems, the flow curves obtained are dependent on the previous shear history of the system.

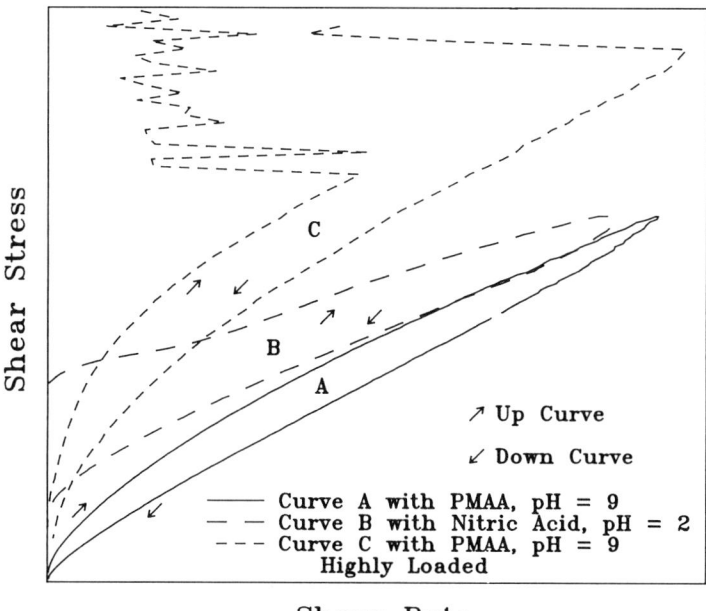

Figure 1 Representative flow curves for alumina slips.

Pseudoplastic behavior with a yield point is illustrated by Curve A. This is a typical curve obtained with the PMAA at low volume fraction alumina loadings. Curve B illustrates the change in flow behavior when nitric acid is the dispersant. The curve has Bingham behavior with a yield point. The PMAA imparts the pseudoplastic behavior to the dispersion. Curve C shows the shear hardening effect or shear thickening effect obtained as the volume fraction loading is increased. The curve follows the pseudoplastic outline until a critical shear rate is reached. Above the critical shear rate the curve exhibits shear fracture. In decreasing stress mode, the normal pseudoplastic behavior is again obtained below a critical shear rate. The transition to a dilatant system is sharp. The stress difference between the yield point and the critical shear stress before dilatancy occurred became narrow as alumina loading increased. This decreasing stress difference is one possible measure of the transition to dilatancy.

In Figure 2 the yield stresses for each slip are plotted. As the volume percent increases with the PMAA dispersions, a sharp rise in yield stress occurs at the point where the materials become dilatant. The higher surface area alumina has a yield stress increase at a lower loading. Acid dispersions have a significantly higher yield stress at lower loadings.

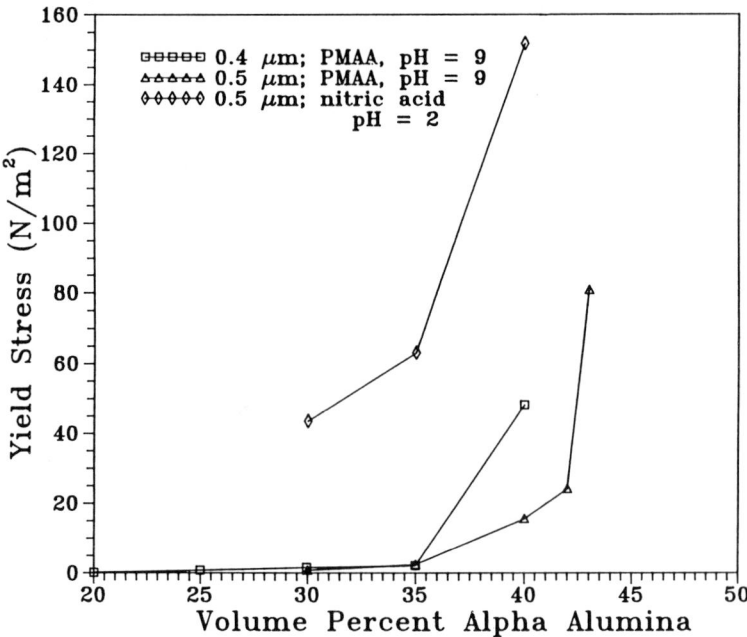

Figure 2 Yield stresses for alumina slips.

The slips were also studied using oscillatory measurements. For this procedure, a sweep of applied torque was made at a constant frequency of 1 Hz. This was done to locate the linear viscoelastic region for each dispersion. The linear viscoelastic region is defined as the range of applied stress where the storage modulus (G′) is nearly constant, the dynamic viscosity (η′) is nearly constant, and the displacement of the measuring head is proportional to the applied stress. Displacements were kept below the level where inertial effects of the measuring head become a factor. Once the linear viscoelastic region was located, the G′ and η′ were averaged. The phase angle was nearly constant, and was also averaged.

Results for the storage modulus are given in Figure 3. For the PMAA dispersions there is a significant jump in elasticity as the volume fraction increases. This corresponds well to regions where shear hardening happens. The higher surface area alumina had a larger storage modulus increase at lower volume fractions. For the acid system the elasticity was significantly higher at all loadings indicating more structure in the system than with the PMAA.

The viscosity component shown in Figure 4 behaves similarly to the storage modulus. The dynamic viscosity builds by roughly an order of magnitude in the region

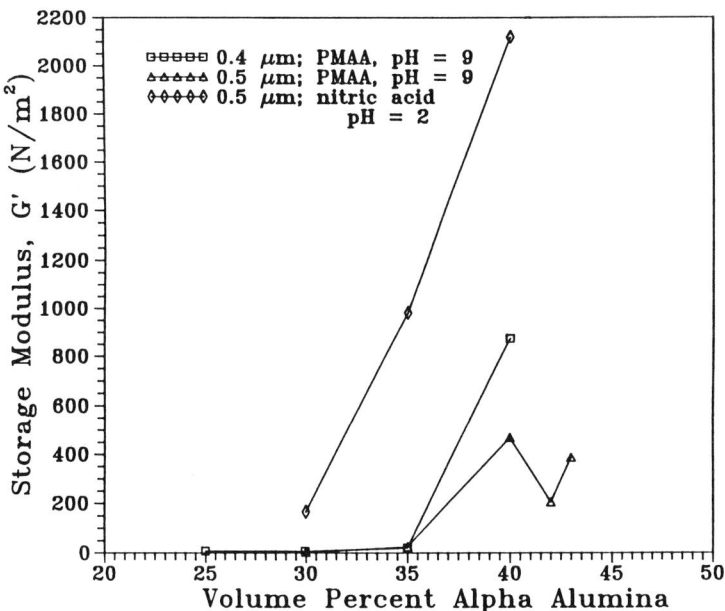

Figure 3 Storage modulus for alumina slips.

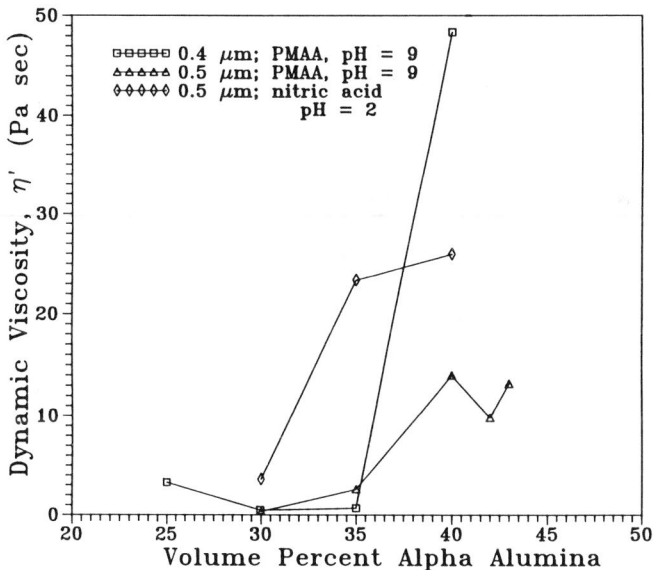

Figure 4 Dynamic viscosity for alumina slips.

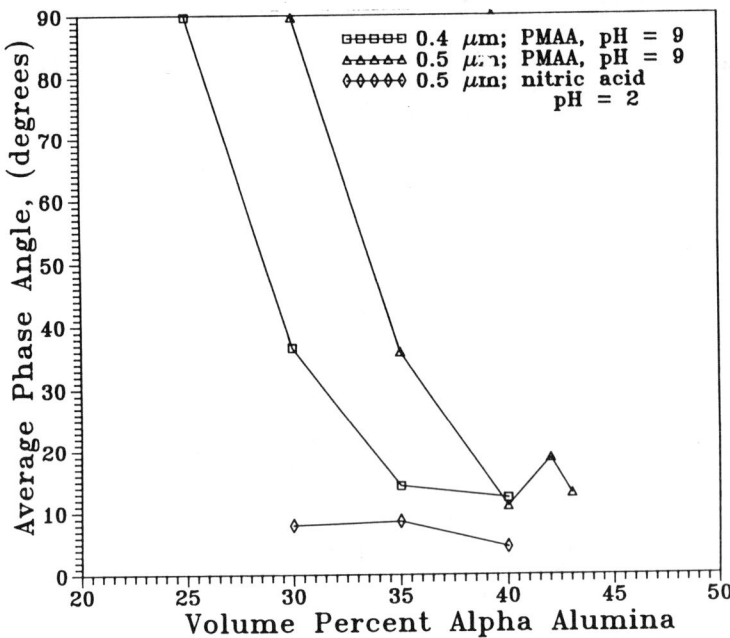

Figure 5 Average phase angle for alumina slips.

where dilatancy occurs. For the acid dispersion the viscosity builds at much lower concentrations.

For oscillatory measurements, the phase angle is the difference between the applied sine wave and the response of the system. For a Newtonian fluid, the phase angle is 90°, and for a solid the phase angle is 0°. Viscoelastic systems are somewhere between, depending on how much structure is present.

Figure 5 shows the average phase angles for the dispersions. For the PMAA dispersed systems, the phase angle drops smoothly with an increase of alumina loading. The lower surface area alumina always has a higher phase angle, indicating less structure in its system. It can be taken to a higher volume percent loading than the other alumina. The acid system phase angle is always very low indicating significant structure even at low volume percents.

CONCLUSIONS

1. Combined flow and oscillatory measurements on alumina slips are useful for assessing the structure within the slip.

2. Transition from a pseudoplastic dispersion to a dilatant dispersion is indicated by a large increase in elasticity, G', and yield strength of the slip. The corresponding phase angle from oscillatory measurements approaches $0°$.

3. The transition of an alumina slip to dilatant behavior is not sharp. As volume fraction alumina increases, dilatancy appears at a smaller applied stress.

REFERENCES

1. I. A. Aksay, "Microstructural Control Through Colloidal Consolidation", pp. 94-104 in Forming of Ceramics (Adv. in Ceramics Vol. 9), ed. J. A. Mangels and G. L. Messing, Am. Ceram. Soc., Columbus, OH (1984).
2. I. A. Aksay, W. Y. Shih and M. Sarikaya, "Colloidal Processing of Ceramics with Ultrafine Particles", pp. 393-406 in Ultrastructure Processing of Advanced Ceramics, ed. J. D. Mackenzie and D. R. Ulrich, John Wiley & Sons, New York (1988).
3. I. A. Aksay, "Principles of Ceramic Shape-Forming with Powder Systems", Ceramica Trans., 1B (1988) pp. 663-674.
4. G. Y. Onoda, "The Rheology of Organic Binder Solutions", pp. 235-251 in Ceramic Processing Before Firing, ed. G. Y Onoda and L. L. Hench, John Wiley & Sons, New York (1978).
5. J. S. Reed, Intro. to the Principles of Ceramic Processing, John Wiley & Sons (1988) pp. 227-250.
6. G. W. Phelps and M. G. McLaren, "Particle-Size Distribution and Slip Properties", pp. 211-225 in Ceramic Processing Before Firing, ed. G. Y. Onoda and L. L. Hench, John Wiley & Sons, New York (1978).
7. M. D. Sacks, "Rheological Science in Ceramic Processing", pp. 522-538 in Science of Ceramic Chemical Processing, ed. L. L. Hench and D. R. Ulrich, John Wiley & Sons (1986).
8. C. F. Zukoski, J. W. Goodwin, R. W. Hughes and S. J. Partridge, "Viscoelastic Probes of Suspension Structure", pp. 231-236 in Better Ceramics Through Chemistry II, Symp. Proc. vol. 73, ed. C. J. Brinker, D. E. Clark and D. R. Ulrich, Mat. Res. Soc., Pittsburgh, PA (1986).
9. H. Okamoto, M. Hashiba, K. Hiramatsu and Y. Nurishi, "Selection of Carboxylates as Dispersants for Alumina in Slip Casting", pp. 224-229 in Sintering 87, ed. S. Somiya, M. Shimada, M. Yoshimura and R. Watanabe, Elsevier New York (1988).
10. J. Cesarano, I. A. Aksay and A. Bleier, "Stability of Aqueous Alpha-Alumina Suspensions with Poly (methacrylic acid) Polyelectrolyte", J. Am. Ceram. Sc., 71(4) pp. 250-255 (1988).

DRYING STRESSES IN GRANULAR CERAMIC FILMS

Raymond C. Chiu and Michael J. Cima
Ceramics Processing Research Laboratory, Massachusetts Institute of Technology,
Cambridge, MA 02139

ABSTRACT

Stress developed in granular alumina films during drying was measured by a substrate deflection method using an optical interference technique. The particle size dependence of the maximum stress was measured by using alumina dispersions containing 0.4 µm or 0.65 µm average size particles. No cracks were observed in films with thickness below 60 µm when prepared from the 0.4 µm slurry. The critical thickness was independent of drying rate over the entire range studied. For thicker films, cracking in the films was found to occur at saturation levels just below 100% for drying rates $>1x10^{-5}$ g/s and at 30-40% for drying rates ten times slower. For films dried at an even slower rate, $6x10^{-7}$ g/s, no cracking was observed.

INTRODUCTION

Liquid vehicles are essential to many ceramic forming processes. Liquid removal prior to firing, however, often causes problems with dimensional control, segregation, and cracking. Addition of binders can alleviate some of the drying defects by strengthening the greenware. Unfortunately, the use of organic processing aids is associated with many different types of defects such as carbon contamination and increased cycle time for component manufacture. Basic information concerning the relationship between drying stress and processing parameters could provide guidance for improving drying conditions or in the design of processing aids to decrease the tendency toward cracking.

Here we report on the development of stress and the formation of cracks in granular ceramic films during drying. An optical interference technique was used to measure stress *in situ* while the mass of the drying film was measured simultaneously.

EXPERIMENTAL PROCEDURE

The granular ceramic films used in this study were made from 20 vol% aqueous dispersions of alumina. The average particle sizes of the powders used were 0.4 μm^* and 0.65 μm.[†] The particulate suspensions were stabilized by adjusting the pH of the dispersions to 3.5 with nitric acid followed by ultrasonication and readjustment to pH 3.5.

A substrate curvature technique was used to measure indirectly the stress produced in the film during drying. Optical interference was used to determine the curvature of a substrate. This method involves counting and measuring the relative positions of the interference rings (Newton's rings) generated by the displacement between the sample and an optical flat when viewed under a monochromatic light source.[1] Since each interference ring represents an additional constant amount of displacement of one-half the wavelength of the light source, the curvature can be calculated from the displacement versus position data. From the curvature, the stress in the film can be determined by $\sigma = Et_s^2/[6(1-\upsilon)Rt_f^2$ where E and υ are the Young's modulus and the Poisson ratio of the substrate, respectively, t_s is the substrate thickness, t_f is the film thickness, and R is the radius of curvature of the substrate.

Films were cast onto (001) silicon substrates of 100 or 130 μm thickness for stress measurements. As received, the Si substrates[‡] were 0.051 m in diameter and were flat to 3 μm. These were cut into rectangular pieces of various sizes. The appropriate substrate thickness was selected to maximize the sensitivity for the stress range measured. Our calculations used established values for the Young's modulus and Poisson ratio of single crystal silicon and were 1.689×10^{11} Pa and 0.064, respectively.[3]

A schematic of the experimental setup for stress measurement is shown in Figure 1. The film was cast onto a silicon substrate which was placed on an optical flat. The optical flat was then hung from an analytical balance.[§] The weight loss during drying was recorded through an interface to a personal computer. Monochromatic light of 0.5461 μm was used to generate interference patterns between the reflective back side of the silicon substrate and the optical flat. Changes in the interference pattern during drying were captured in real time using a video camera and video recorder while simultaneously recording the weight. The desired frames of the video recording were captured by a video frame digitizer for processing after the completion of drying. The curvature of the substrate was determined at selected time intervals by analyzing the positions and number of fringes in the digitized image. Calculation of the average stress in the film was performed using the measured curvature at different times during drying. Since even a small air current could disrupt the mass measurements, a shield was placed around the setup to minimize disturbance. For the same reason, ambient drying conditions were not

---

*   AKP-30, Sumitomo Chemical Co., Tokyo, Japan.
†   AKP-15, Sumitomo Chemical Co., Tokyo, Japan.
‡   Virginia Semiconductor, Fredericksburg, VA.
§   Model AE160, Mettler Instrument Corp., Hightstown, NJ.

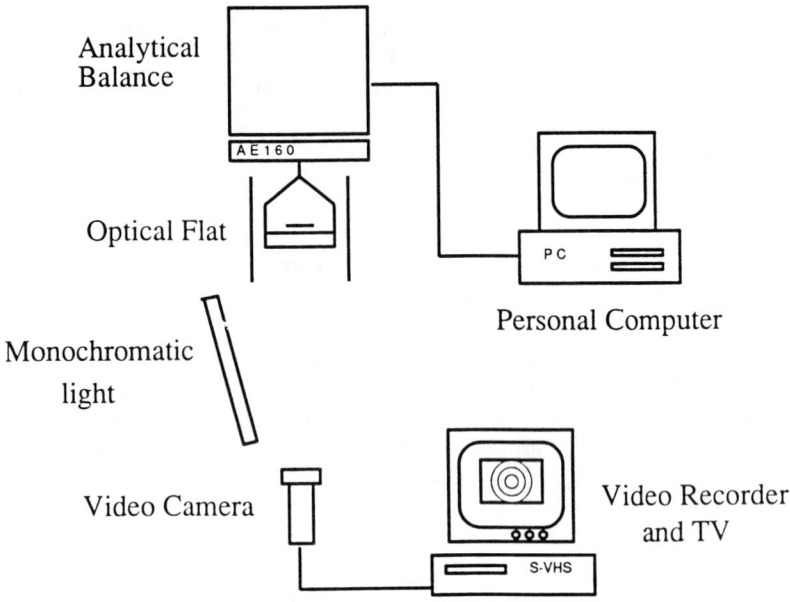

Figure 1 Schematic of experimental setup for stress measurement.

controlled but were monitored using a solid state hygrometer and a thermocouple. The drying rate, however, could be reduced by placing a cover with a small opening over the film during drying.

Mercury porosimetry[*] was used to measure the dried density of the green films. Due to the volume of sample required for this, a number of films dried under the same conditions were mechanically removed from the substrate and collected for the measurement.

RESULTS AND DISCUSSION

During the drying of these films, supersaturated, saturated, and dry regions could be observed. In the supersaturated region, the film was still fluid. Under an optical microscope, the saturated region appeared translucent in transmission mode. In the dry region, the films appeared opaque. An optical micrograph showing the three regions during drying is shown in Figure 2. For all films used in the stress measurements, only the saturated and supersaturated regions coexisted during drying.

---

[*] Autopore II, Micromeritics Instrument Corp., Norcross, GA.

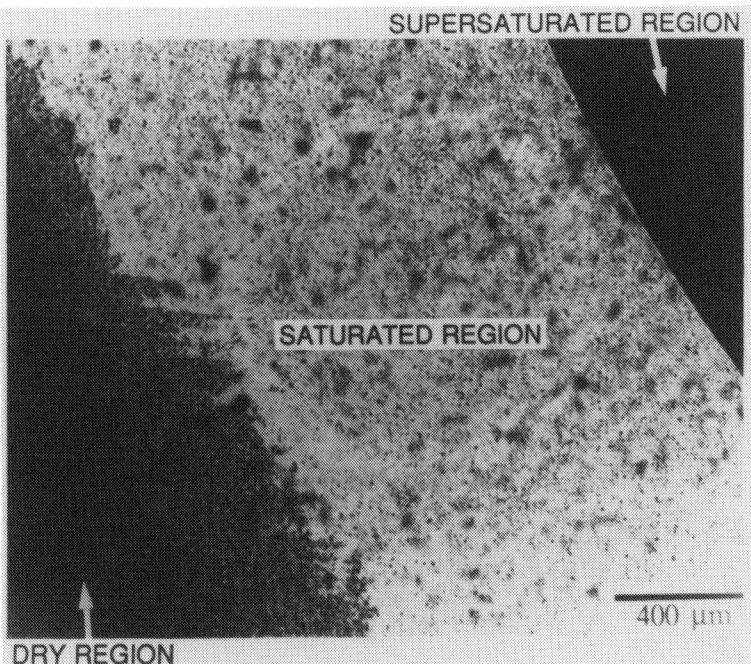

Figure 2 Optical micrograph in transmission mode of a drying film. The supersaturated region is to the right, the dry region is to the left, and the translucent saturated region lies between the two.

Shown in Figure 3 are the stress histories for 40 µm thick films produced from the 0.4 µm and 0.65 µm slips cast onto 0.009 x 0.009 m^2, 100 µm thick Si substrates. The surface evaporative flux for these films was measured to be 7.2x10^{-6} kg/m^2-s. The stress did not develop until the extinction of the supersaturated region, which is indicated in Figure 3 by the vertical dashed line. Thereafter, the stress rapidly reached a maximum and then decreased monotonically. Under identical drying conditions, the film produced from the 0.4 µm average particle size dispersion took twice the time to reach maximum stress than did that from the 0.65 µm slip.

For films produced from the 0.4 µm slip having thicknesses >60 µm, cracking in the films occurred during drying. The point at which cracking occurred can be determined accurately since a distortion in the optical interference pattern due to the cracking can be detected. For 80 µm thick films produced from the 0.4 µm average particle size slip, the cracking behavior was found to be dependent on the external drying conditions. At a drying rate of 1x10^{-8} kg/s, extensive cracking and spalling was observed, and cracking always occurred at saturation levels just after the

91

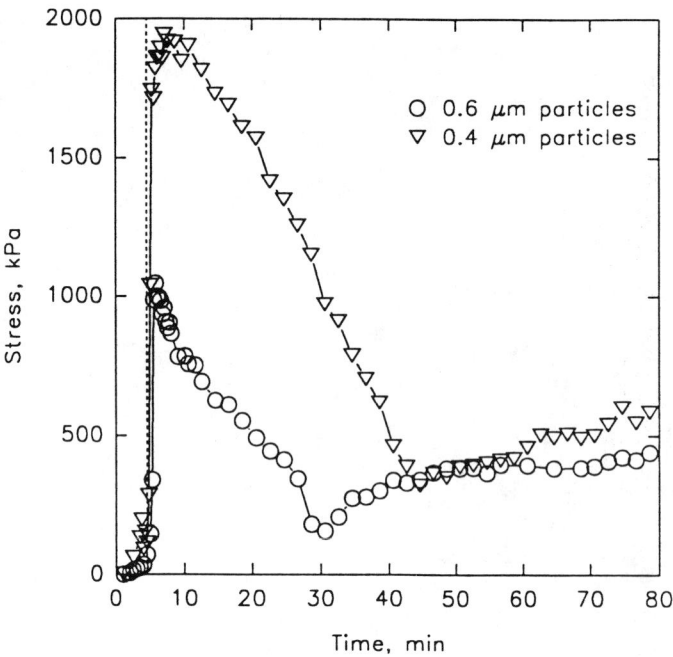

Figure 3 The stress histories for 40 μm thick films produced from the 0.4 μm and 0.65 μm average particle size dispersions. The substrate was 0.009 x 0.009 m² and 100 μm thick.

measured stress reached a maximum. When the drying rate was reduced to a lower value (2×10^{-9} kg/s), cracking was found to occur between 30-40% saturation with only a few cracks forming. At a drying rate of 6×10^{-10} kg/s, no cracks were observed in the films. Mercury porosimetry measurements indicated the porosities of all of the dried films were 62% of theoretical.

In a previous report, a length scale for film dimensions was established within which capillary flow could act and ensure uniform saturation during drying.[4] For the external drying rates used, the dimensions of films produced in this study were well within the estimated length of approximately 1×10^{-2} m for the saturated region. Therefore, only the supersaturated and saturated regions were observed to coexist during drying.

As observed under the optical microscope, the saturated region appeared translucent and so was assumed to be fully saturated. As drying proceeded, the films turned opaque as pores began to empty and light scattering increased. Since the

92

capillary force was acting to redistribute liquid in the saturated region, the absence of any measured average stress until the supersaturated region disappeared indicated the film must not have been rigid in the saturated state. This suggested that the particles in the saturated region were still separated by an electrostatic repulsive layer in which water between the particles provided very little shear resistance to sustain any stress. The stress development following the supersaturated region extinction indicated the film was becoming rigid and was responding to the stress imposed by the capillary redistribution pressure. A comparison of the hydrostatic compressive force imposed by the capillary pressure and the electrostatic repulsive force between the dispersed particles could determine whether the particles would touch. From Monte Carlo simulation, Van Megen et al. determined that the modulus of a model colloidal network having particle-particle interaction based on classical DLVO potential is on the order of 10^2-10^3 Pa for submicrometer particles.[5] Given that the measured stress was over 1 MPa, the particles should touch as capillary forces act. When the particles touched, the film behaved elastically, and the measured average stress reflected the capillary pressure acting in the pores. Correspondingly, as the population of pores decreased during drying, the average stress measured decreased.

The thickness of the electrostatic repulsive layer between dispersed particles can be estimated by the inverse of the Debye-Hukel parameter, κ. At room temperature in an aqueous medium, $\kappa = 3.288 I^{0.5}$ (nm)$^{-1}$ where I is the ionic strength in mol/L.[6] At a pH of 3.5, the Debye layer was estimated to be 8 nm thick. Since the Debye layer thickness relative to the particle size is larger for smaller particles, the amount of water lost between the supersaturated region extinction state and the stress maximum state should be larger for films produced from smaller particles. Packing densities in the saturated state can be calculated based upon the dry film weights and the amounts of water lost during drying. Indeed, films produced from 0.65 μm particle size slips had a wet density of 60% of theoretical, and those produced from 0.4 μm particles had a wet density of 57% of theoretical. The error in the calculated densities is 1%. Since both films had dried green densities of 62% of theoretical, the films produced from smaller particles must have retained more liquid in the saturated state, which is consistent with our discussion.

The cracking behavior dependence on drying rate suggested that cracking was due to two different states of stress. At high saturation, shrinkage in the films occurred as the particles were brought together by capillary forces. The point where cracking occurred coincided with the point where the shrinkage was expected to cease in the films. The behavior is similar to that described for the drying of gels where failure is typically observed as soon as shrinkage stops.[7,8] In these cases, failure is avoided at extremely slow drying rates.

Associated with the negative suction pressure is a biaxial tensile stress on the material surrounding the pores. In the latter stage of drying, when the remaining fine pores begin to empty, a state of nonhomogeneous biaxial stress exists in the film. Therefore, failure could occur during the latter stage of drying even at drying rates slow enough to avoid cracking in the early stage. Unfortunately, the stress measurement setup used in this study can only measure the average stress in the films during drying. Therefore, nonhomogeneous stress distribution, which may be responsible for failure, cannot be assessed.

CONCLUSIONS

Stress measurements by a substrate deflection method were used to measure the average stresses in granular ceramic films. From the simultaneous stress and weight loss measurements, the point where the particles in the green films touched was estimated. The shrinkage due to the collapse of the Debye layer occurred in an early stage of drying when capillary forces act. With the experimental setup, the point where cracking occurred in these films was found to be dependent on the drying rate. The cracking that occurred in the early stage of drying was believed to be similar to that observed in gels. In the latter stage of drying, the nonhomogeneous distribution of capillary stress is believed to be responsible for the cracks observed.

REFERENCES

1. J.D. Finegan and R.W. Hoffman, "Stress and Stress Anisotropy in Iron Films"; pp. 935-42 in the Proceedings of the Eighth National Symposium on Vacuum Technology Transactions. Pergamon Press, New York, NY, 1961.
2. R.W. Hoffman, "Mechanical Properties of Thin Condensed Films"; pp. 211-73 in Physics of Thin Films. Edited by G. Hass and R.E. Thun. Academic Press, New York, NY, 1966.
3. W.A. Brantley, "Calculated Elastic Constants for Stress Problems Associated with Semiconductor Devices," *J. Appl. Phys.*, **44** [1] 534-35 (1973).
4. R.C. Chiu and M.J. Cima, "Drying Behavior of Granular Ceramic Films"; in Ceramic Powder Science IV. Edited by G.L. Messing. American Ceramic Society, Westerville, OH, in press.
5. W.J. Van Megen, I.K. Snook, and R.O. Watts, "Elastic Properties of Model Colloids," *J. Colloid Interface Sci.*, **77** [1] 131-36 (1980).
6. R.J. Hunter, Zeta Potential in Colloid Science; p. 27. Academic Press, San Diego, CA (1980).
7. R.K. Dwivedi, "Drying Behavior of Alumina Gels," *J. Mater. Sci. Lett.*, **5**, 373-76 (1986).
8. P. Anderson and L.C. Klein, "Shrinkage of Lithium Aluminosilicate Gels during Drying," *J. Noncryst. Solids*, **93** 415-22 (1987).

EFFECTIVE PERCOLATION LIMIT IN THE DRYING OF CERAMIC COMPOSITES

Martin W. Weiser and Kami C. Key
Mechanical Engineering Department and UNM/NSF Center for
Micro-Engineered Ceramics, University of New Mexico, Albuquerque, NM 87131

ABSTRACT

The green density of slip cast ceramic composites was found to be strongly dependent upon whether the volume fraction of inclusions was above or below the effective percolation limit for the system. This effective percolation limit was found to differ from the theoretical value which is based strictly upon the inclusion morphology. The value of the effective percolation limit was found to depend upon the capillary forces between the matrix particles and the degree of lubrication between particles in addition to the inclusion morphology.

INTRODUCTION

The need to fabricate higher quality ceramic composites has lead to numerous investigations of methods to improve the green homogeneity and final grain size of the materials. Many of these investigations have concentrated on using finer powders and dispersants to improve both the distribution of the various phases in the green body and the densification of the compact upon firing. As the particle sizes have decreased below 0.5 μm the problem of cracking during drying has resurfaced.[1] This is particularly evident in the drying of gels where it is very difficult, if not impossible, to form large monolithic bodies.

During the drying of a powder compact internal stresses develop due to differential shrinkage between various regions of the compact.[2-5] These shrinkage stresses arise as a result of the capillary forces between powder particles and the formation of stiff, dry regions adjacent to compliant, wet regions due to the progression of the drying front.[6] This process is very similar to the development of flaws during sintering.[7] The capillary forces between particles are approximately inversely proportional to the particle size so they can become very large for small particles resulting in localized fracture of the drying compact. In addition, the drying rate of the compact decreases with decreasing particle size because the pore sizes and the mean free path for diffusion decrease. This effect can be partially countered by drying the compacts at higher temperature and/or higher humidity to improve the diffusion or flow rate of the solvent and to control surface evaporation.[8] The use of higher drying temperatures and humidity can be extended to include critical point drying where the liquid to vapor phase transition is eliminated.

The presence of inclusions that are larger than the matrix particle size such as agglomerates and reinforcing particles or whiskers can result in the formation of a stiff network that can not be deformed to accommodate the shrinkage of the matrix. This will result in the formation of flaws in the matrix particle packing which will be difficult to eliminate during sintering. The effects of such a network are seen most frequently when the percolation limit for inclusion contact is exceeded.[9]

A number of different models of the percolation limit have been developed which differ in primarily in their assumptions of nature and distribution of the percolating phase. The classical case is for hard spheres that are randomly distributed on a lattice which yields a percolation limit of 16 vol%. Later studies examined random distributions of hard spheres, soft-shelled spheres, and both hard and soft shelled rod-shaped particles.[10,11] The percolation limit for randomly distributed hard spheres in a continuous matrix has been found to be as high as 64 vol%, although lower values are expected in most practical cases. The percolation limit of soft-shelled particles was found to be less than that of hard particles and depended upon the thickness of the shell. This is consistent with the findings of a previous study of the effect of inclusion size on densification[12] in which a shell of matrix particles was found to form around inclusions such that smaller inclusions impeded densification more than larger ones at the same volume fraction.

The actual percolation limit for a particular inclusion morphology may not be the most important item for determining the homogeneity of the green compact; it may be an effective percolation limit that can be either lower or higher than the predicted value. The effective value would be lower as a result of lubrication between inclusions[13] or the formation of a soft-shell of matrix particles surrounding the inclusion. It would be higher as a result of the capillary forces between the matrix particles which act to collapse and rearrange the inclusion network.

EXPERIMENTAL PROCEDURE

Composites composed of fine Al_2O_3 powder[*] which contained various volume fractions of larger SiC[†] and ZrO_2[‡] inclusions were fabricated by slip casting. The

---

[*] AKP-15, 30, and 50 supplied by Sumitomo Chemical Co.
[†] Supplied by Norton Co. and air classified in a previous study.[12]
[‡] E30, SC30, and SC101 supplied by Magnesium Elektron and DK-2 supplied by Zirconia Sales (America) Inc.

slips contained 0 to 60 vol% inclusions and were dispersed using Na-PAA (SiC inclusions) and Na-PMAA* (ZrO₂ inclusions). The slips were then ultrasonically disrupted or vibratory milled for 10 minutes to obtain homogeneous slips of near minimal water content. These slips were then cast on the surface of a dampened plaster bat in cylindrical plastic molds that were either 12 or 30 mm in diameter. The pellets were dried on the plaster for ≈24 hours followed by ≈24 hours at 85°C to remove the remaining water. After cooling to room temperature, the density of each pellet was determined from measurement of the mass, diameter, and thickness of the pellet.

RESULTS AND DISCUSSION

The fractional theoretical densities of a series of compacts made from AKP-15 Al₂O₃ (0.68 μm) and 10 μm SiC powder which had been air classified are shown in figure 1. The bulk density includes the mass and volume of both the matrix the inclusions while the matrix density excludes the inclusion mass and density. There is a dramatic break in the density as a function of inclusion volume fraction at approximately 27 vol% SiC. This break indicates that the mechanism controlling the

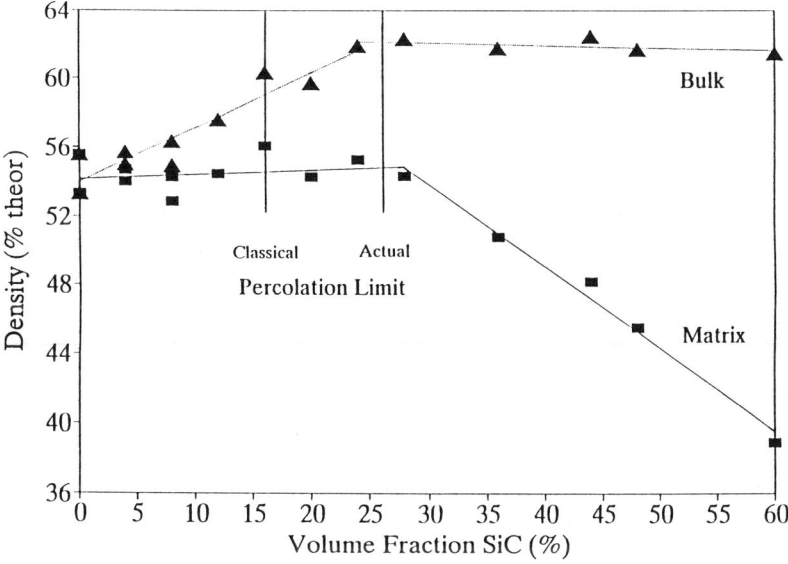

Figure 1 The bulk and matrix fractional densities for composites composed of 10 μm SiC in a 0.68 μm Al₂O₃ matrix. The predicted percolation limit is based upon randomly distributed spherical inclusions while the actual percolation limit is derived from the break in the density versus volume fraction of inclusion curves.

---

* Daxad 30 and 37 LN10-35 supplied by W.R. Grace & Co.

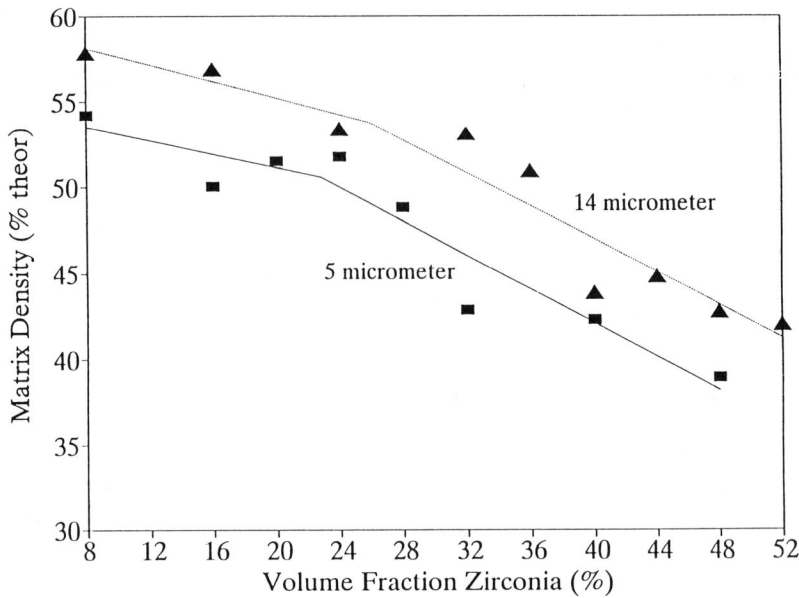

Figure 2 The fractional matrix density of two series of composites made from 5 and 14 μm ZrO_2 inclusions in a 0.68 μm Al_2O_3 matrix.

density of the compact is changing from packing of the matrix particles at low volume fractions of inclusions to the packing of inclusion particles at high volume fractions.

The matrix densities for two different series of compacts made from AKP-15 (0.68 μm) Al_2O_3 and 5 or 14 μm ZrO_2 particles are shown in figure 2. The effective percolation limit and density for the composite containing 14 μm ZrO_2 particles are somewhat higher than those for the composite containing 5 μm particles. The higher percolation limit is also observed in the composite made with the larger inclusions and a finer matrix powder (AKP-30, 0.38 μm) as shown in figure 3. However, in this case the composites containing 5 μm ZrO_2 particles were more dense.

The effective percolation limit is higher than the classical percolation limit based upon the inclusion morphology as a result of the capillary forces between particles during drying and the random distribution of the inclusions. The capillary forces tend to collapse the inclusion network that develops as the inclusion concentration reaches the percolation limit. The effective percolation limit is higher for the composites made with the finer AKP-30 matrix than for the coarser AKP-15 matrix.

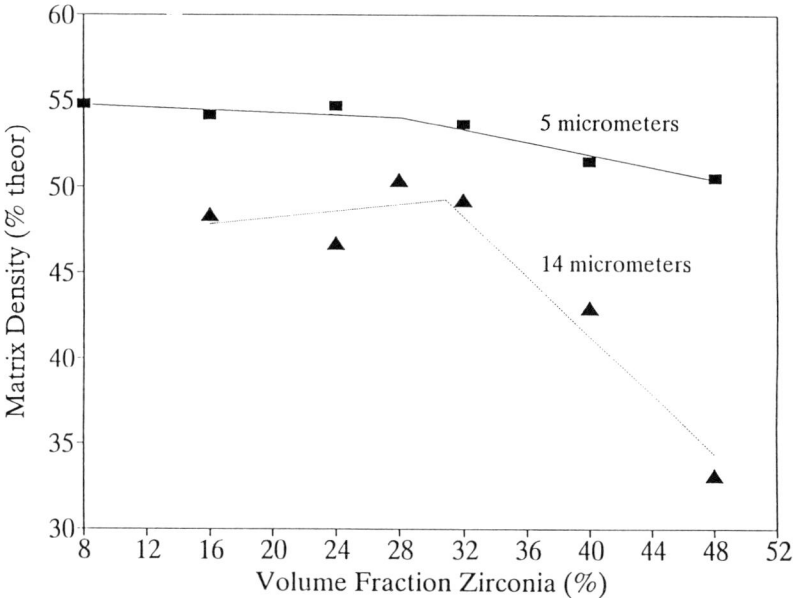

Figure 3 The fractional matrix density of two series of composites made from 5 and 14 μm ZrO$_2$ inclusions in a 0.39 μm Al$_2$O$_3$ matrix.

This is the expected behavior since the capillary forces during drying are higher in the finer matrix. The effective percolation limit is lower for the 5 μm inclusions in both the AKP-15 and AKP-30 matrices. This is a consequence of the more numerous interparticle contacts for the smaller inclusions. The more numerous interparticle contacts increase the frictional forces between inclusions and limits rearrangement of the matrix particles making it more difficult to deform the percolation network.

CONCLUSIONS

The green density of slip cast ceramic matrix composites containing equiaxed inclusions has been studied as a function of volume fraction inclusions, matrix particle size, and inclusion particle size. It was found that the density changed behavior as a function of inclusion volume fraction at 22 to 30 vol% inclusions. This change of behavior is attributed to the formation of a percolation network by the inclusions that impedes further shrinkage. This effective percolation limit is higher than the classical percolation limit based upon spherical inclusions. It was found that the effective percolation limit was higher for larger inclusions and finer matrices which is consistent with the percolation network being collapsed by the capillary forces during drying.

ACKNOWLEDGMENTS

The authors would like to thank W. David Willis for assistance in preparing the initial set of Al_2O_3/SiC composites described in figure 1 and G.W. Scherer and M.J. Cima for useful discussions. This work was supported by the Air Force Office of Scientific Research under contract 90-0265 and Sandia National Laboratories under contract 54-9308.

REFERENCES

1. G.W. Scherer, "Theory of Drying", *J. Am. Ceram. Soc.*, **73** [1] 3-14 (1990)
2. R.Q. Packard, "Moisture Stress in Unfired Ceramic Clay Bodies", *J. Am. Ceram. Soc.*, **50** [5] 223-29 (1967)
3. H.H. Macey, "Clay-Water Relationships and the Internal Mechanism of Drying", *Trans. Br. Ceram. Soc.*, **41** [4] 73-121 (1942)
4. K.J. Packer, "The Dynamics of Water in Heterogeneous Systems", *Philos. Trans. R. Soc. London, Ser. B*, **278** 59-87 (1977)
5. A.R. Cooper, "Quantitative Theory of Cracking and Warping during the Drying of Clay Bodies", pp 261-76 in <u>Ceramics Processing Before Firing</u> ed. by G.Y. Onada Jr. and L.L. Hench, Wiley, New York, 1978
6. T.M. Shaw, "Movement of a Drying Front in a Porous Media", pp 215-23 in <u>Better Ceramics Through Chemistry II</u> ed. by C.J. Brinker, D.E. Clark, and D.R. Ulrich, Materials Research Society, Pittsburgh, PA 1986
7. M.W. Weiser and L.C. De Jonghe, "Rearrangement During Sintering in Two-Dimensional Arrays", *J. Am. Ceram. Soc.*, **69** [11] 822-26 (1986)
8. T. Mizuno, H. Nagata, and S. Manabe, "Attempts to Avoid Cracks during Drying", *J. Non-Cryst. Solids*, **100** 236-40 (1988)
9. F.F. Lange, "Constrained Network Model for Predicting Densification Behavior of Composite Powders", *J. Mater. Res.*, **2** [1] 59-65 (1987)
10. I. Balberg and N. Binenbaum, "Scher and Zallen Criterion: Applicability to Composite Systems", *Phys. Rev. B*, **35** [16] 8749-52 (1987)
11. E.A. Holm and M.J. Cima, "Zero-Shrinkage Whisker Fraction in Ceramic Matrix-Ceramic Whisker Composites", pp 319-30 in <u>Processing Science of Advanced Ceramics</u> ed. by I.A. Aksay, G.L. McVay, D.R. Ulrich 1989
12. M.W. Weiser and L.C. De Jonghe, "Inclusion Size and Sintering of Composite Powders", *J. Am. Ceram. Soc.*, **71** [3] C125-C127 (1988)
13. G.W. Scherer, "Viscous Sintering of Particle-Filled Composites", *Am. Ceram. Soc. Bull.*, **70** [6] 1059-63 (1991)

DRYING OF GELCAST CERAMICS

Ogbemi O. Omatete, Richard A. Strehlow and Claudia A. Walls
Oak Ridge National Laboratory, Oak Ridge, TN

ABSTRACT

Plates (8.9 x 90.2 x 228.6 mm) of silicon nitride, sialon, and alumina made by gelcasting, a generic near-net-shape forming process, have been dried in a controlled humidity chamber. No constant-rate period was found in drying the gelcast plates. A linear shrinkage of 3-4% was observed, the shrinkage stopping after only about 20% of the total moisture had been lost. Dried parts, free of warpage and cracking, were obtained in reasonable times by controlling the humidity and temperature.

INTRODUCTION

Gelcasting[1,2] is a generic, near-net-shape forming process which is applicable to complex-shaped ceramic parts. In the process, free-flowing concentrated suspensions (>50 vol% solids) of commercial powder in a solution of organic monomers, are cast and polymerized to form a filled gel which takes the shape of the mold. The gelcasting of alumina powders[3] and of sialon and silicon nitride powders[4,5] have been reported. The process has also been successfully applied to composites.[6,7]

Gelcasting should be distinguished from the sol-gel process. Gelcasting utilizes high solids loading of commercial powders in a solution of organic monomers that are subsequently polymerized; whereas, a sol-gel process typically produces its solids as part of the process in the form of an inorganic gel at very low solids loading. Gelcasting has several advantages over the existing complex shape forming methods such as slip casting and injection molding.[3-7] Among these, are its simplicity and overall processing speed. However, we have found it necessary to dry the parts at high relative humidities (80-96%). Consequently, the drying of thick parts (>15 mm) may require over a hundred hours.[3] To maintain the speed of the process and optimize process economy, it is essential to reduce the drying time.

Drying is a unit operation that has been thoroughly studied[8,9] and is a well-known part of ceramic processing.[10] An exhaustive treatment of the drying of gels in sol-gel processing has recently been published.[11,12] The primary objective of this work is to study the drying of gelcast parts with a view to minimizing the time needed to produce stress-free, dried bodies while maintaining part-to-part size uniformity.

EXPERIMENTAL PROCEDURE

The powders used in this study were: alumina RCHP-DBM[*] (d_{mean} = 0.6μm), sialon AA[†] (S_{mean} = 9 m^2/g), and silicon nitride, Denka 9S[‡] (d_{mean} = 1.25μm). The gel-casting reagents were the monomer, acrylamide [CH_2=$CHCONH_2$], the crosslinking agent, N,N'-methylene bisacrylamide [$(CH_2$=$CHCONH)_2CH_2$], and the free radical initiator, ammonium persulfate [$(NH_4)_2S_2O_8$]. The detailed process has been described previously.[3-5]

Two drying chambers were built. In one made of plexiglas, the relative humidity was controlled at room temperature. The plate specimen was placed on a wire mesh platform suspended from a balance. Steel rulers with graduations of 0.5 mm were placed along the breadth and length of the specimen, and readings were made periodically using a magnifying lens. In the other, a modified environmental chamber, the relative humidity and the temperature were controlled independently. In both chambers, the variation of the mass of the drying part with time was recorded, and high velocity fans circulated the air in the chamber.

A slurry containing either 51.9 vol% alumina, or 51.2 vol% silicon nitride, or 54.2 vol % sialon was cast in an anodized aluminum mold that produced a plate (8.9 x 90.2 x 228.6 mm). After gelation, the plate was removed from the mold and put immediately into the drying chamber with preset relative humidity and temperature. While it dried, its dimensions were measured and the plate was monitored for cracks and warpage. The plate remained in the chamber until its mass became constant for at least two hours. It was then placed in an oven at 65°C for two hours and its final mass was measured as the bone-dry mass. At room temperature (21-23°C), sialon plates were dried at relative humidities of 75%, 85%, 92% and 96%; humidities for silicon nitride and alumina plates were 75% and 92%. An alumina plate was also dried at 50°C and 75% relative humidity to determine the effect of temperature.

RESULTS AND DISCUSSION

The results of the drying studies are shown in Figs. 1-4. The moisture content X is defined:

---

* Malakoff Industries, Inc., Malakoff, TX.
† Vesuvius Research, Pittsburgh, PA.
‡ Denka, New York, NY.

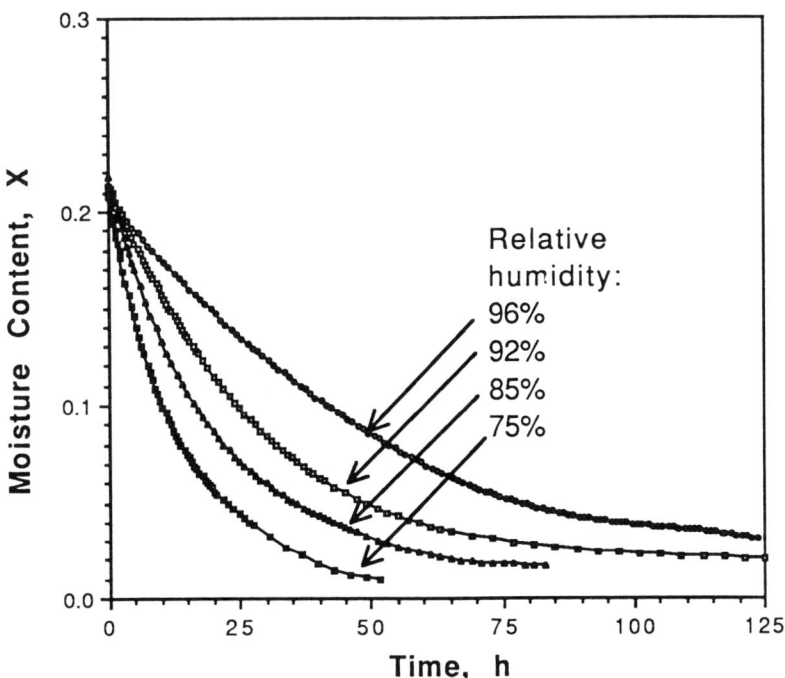

Figure 1 Moisture content vs. time at various relative humidities for sialon at 54.7 vol % and Temp. = 21-23°C.

$$X = [M_{wb} - M_{bd}]/M_{bd} \qquad (1)$$

where M_{wb} is the mass of the wet body and M_{bd} is the bone dry mass. The rate of drying, $-dX/dt$, does not include the area of the drying plate. Figure 1 shows the drying of sialon at several relative humidities. The total time for drying depends on the relative humidity; the lower the relative humidity, the shorter the time of drying.

Figure 2 shows the variation with time of both the moisture content and the percent linear shrinkage of silicon nitride and alumina plates at 92% relative humidity. The shrinkage does not change after about 20 hours although the moisture content continues to change for up to 100 hours. In general, the shrinkage ceased when about 20% of the moisture had been removed irrespective of the relative humidity or material. The rate was controlled by the relative humidity. This can be interpreted as follows. As the body loses moisture, the gel shrinks and the particles move

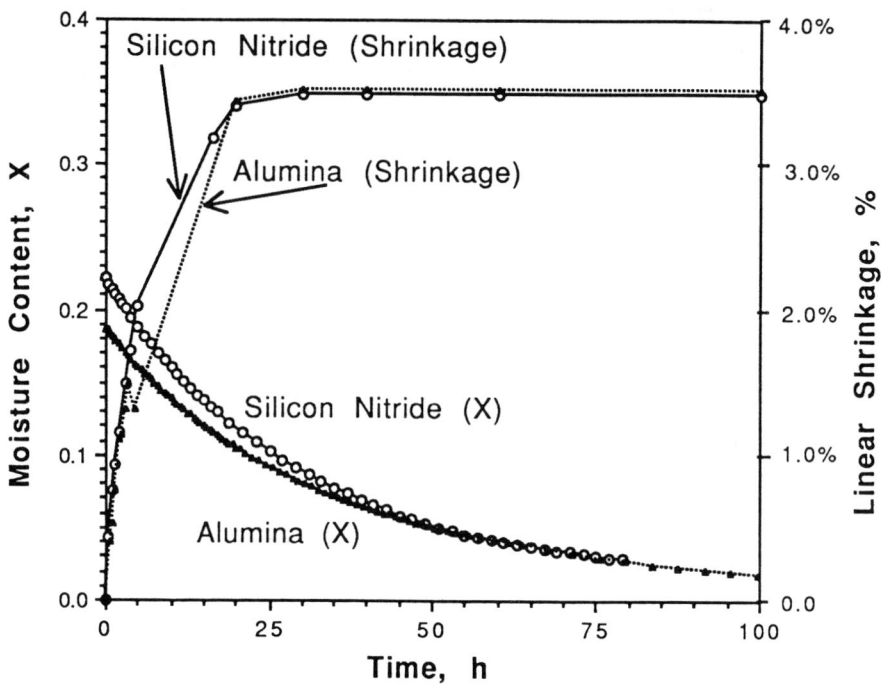

Figure 2 Variation of moisture content and percent linear shrinkage with time (R. H. = 92 %).

closer until there is particle-to-particle contact. Shrinkage of the body then ceases. The gel shrinks and tears as drying continues. Because the particles are bonded by the dried gel, gelcast parts have relatively high strength when dried and are readily machinable.

The tearing of the gel between the fine ceramic powders has the beneficial effect of creating pathways that facilitate the escape of the gases produced during binder burnout. Because shrinkage is limited, we are offered a means to reduce the total drying time. A gelcast part may be dried slowly at high relative humidity with minimal stresses until shrinkage stops. The humidity is then reduced and the drying is completed rapidly at the lower relative humidity.

Figure 3 shows the drying rate curve at various relative humidities for sialon and at 75% relative humidity for alumina and silicon nitride. Unlike typical drying of slips and gels, the figure shows no constant-rate drying period. The absence of a constant-rate period is independent of temperature as seen in Fig. 4 which shows

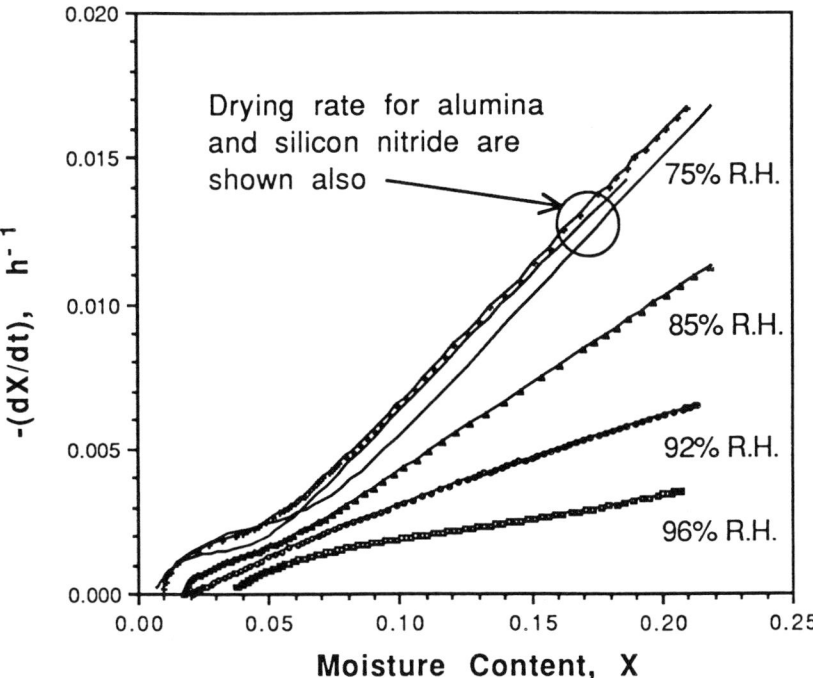

Figure 3 Drying rate curves for sialon at T = 21-23°C show no constant-rate period.

the rate of drying curve for alumina at 75% relative humidity at room temperature and 50°C, respectively. The absence of a constant-rate period means that evaporation is not surface controlled and that the drying is controlled from the beginning by some internal moisture movement. The nature of this internal moisture movement, whether diffusion- or capillary-controlled, will be investigated in the continuing study of gelcast ceramic parts.

Figure 4 shows that the drying rate is increased at a higher drying temperature even at the same relative humidity. This offers a second way to reduce the total drying time. The gelcast part is first dried slowly at high relative humidity and room temperature until shrinking stops and then the temperature may be raised and/or the humidity may be lowered to complete the drying.

The implications and explanations of some of these findings, especially the absence of a constant-rate period, will be the focus of future work. In addition, the effect of the variation of sample shapes and sizes will be examined also.

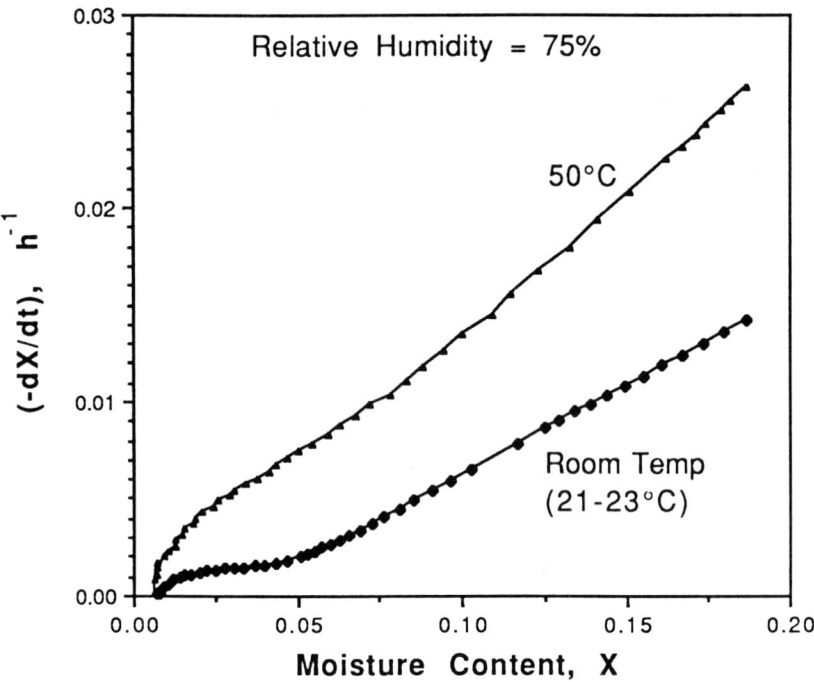

Figure 4 Drying rate curves for alumina at two temperatures

CONCLUSIONS

Drying, a critical step in optimizing the gelcasting process, has been studied and the following were observed:

1. Humidity is the dominant variable and room-temperature drying may be adequate for thin parts.

2. There is no constant-rate drying period.

3. The length of drying can be significantly reduced by varying the humidity and/or the temperature.

ACKNOWLEDGMENTS

Supported by the Ceramic Technology for Advanced Heat Engine Project, Office of Transportation Systems, US Department of Energy under Contract No. DE-AC05-840R21400 with Martin Marietta Energy Systems, Inc.

REFERENCES

1. M. A. Janney, "Method for Forming Ceramic Powders into Complex Shapes," U.S. Pat. No 4 894 194, (Jan. 1990).
2. M. A. Janney and O. O. Omatete, "Method for Molding Ceramic Powders using a Water-based Gelcasting Process," U.S. Patent pending (1991).
3. A. C. Young, O. O. Omatete, M. A, Janney, and P. A. Menchhofer, "Gelcasting of Alumina," J. Am. Ceram. Soc., 74 (3) 612-18 (1991).
4. O. O. Omatete, R. A. Strehlow, and C. A. Walls, "Gelcasting of Submicron Alumina, Sialon and Silicon Nitride Powders," Proc. of the 37th Sagamore Army Materials Research Conference, Plymouth, MA, (1990).
5. O. O. Omatete, R. A. Strehlow, and B. L. Armstrong, "Forming of Silicon Nitride by Gelcasting," Proceedings of the Annual Automotive Technology Development Contractors' Coordination Meeting, P-243, pp 245-251, Society of Automotive Engineers, (1991).
6. O. O. Omatete, T. N. Tiegs, and A. C. Young, "Gelcast Reaction-Bonded Silicon Nitride Composites," Proceedings of the 15th Annual Conference on Composites and Advanced Ceramics, Cocoa Beach, FL (1991)
7. O. O. Omatete, A. Bleier, C. G. Westmoreland, and A. C. Young, "Gelcast Zirconia-Alumina Composites," Proceedings of the 15th Annual Conference on Composites and Advanced Ceramics, Cocoa Beach, FL (1991)
8. H. F. Porter, G. A. Schurr, D. F. Wells, and K. T. Semrau, "Solids Drying and Gas-Solid Systems," Section 20 in Perry's Chemical Engineers' Handbook, 6th edition, R. H. Perry, D. W. Green and J. O. Maloney, ed., McGraw-Hill, NY (1984).
9. R. E. Treybal, "Mass-Transfer Operations," pp 655-717, 3rd edition, McGraw-Hill, NY (1987)
10. J. S. Reed, "Introduction to the Principles of Ceramic Processing," pp 411-426, John Wiley, NY (1988).
11. G. W. Scherer, "Theory of Drying," J. Am. Ceram. Soc., 73 [1] 3-14 (1990).
12. C. J. Brinker and G. W. Scherer, "Sol-Gel Science," pp 407-515, Academic Press (1990).

WHY CRACKS APPEAR WHEN THE CONSTANT RATE PERIOD STOPS IN A DRYING PROCESS

Xiangyue Li
Instituto de Ingenieria, UNAM, Apartado Postal 70-455, 04511,
Mexico D.F., Mexico

ABSTRACT

So far no generally accepted explanation has been established for cracking phenomenon during a drying in gels, cement, concrete and clay processing. In the present work, a fracture mechanics based model is formulated for which special attention is paid on the interaction between fluid and crack propagation mechanism. It has been recognized that, once the constant rate period stops and air begins to enter into pores, owing to flow and diffusion process, the liquids within preexisting cracks disappear making the cracks grow. The present model is capable of including many factors, some of which are evaporation rate, pore size distribution, capillary tension, as well as elasticity, permeability and fracture toughness of solid skeleton material. The theoretical predictions agree with existing experimental observations and it is expected that an optimal design criterion for sol-gel drying process would be established using the present theory.

INTRODUCTION

Unequivocally, it is observed that cracking phenomenon appears to occur in a drying gel body at a so-called critical point (CP). The CP defines the instance when the menisci formed on evaporation surface retrocede into the pores of the body and the surface becomes opaque. In many material, the CP also divides a drying process into two stages: constant rate period (CRP) and falling rate one (FRP). The sudden appearance of cracking at the CP has been reported in many types of gels[1-5] and may be observed also in cement, concrete[6] and clay.[7] Although there exist numerous works in the technical literature aiming in calculating drying-induced stress, no generally accepted explanation for the cracking phenomenon has been established. Two approaches--macroscopic and microscopic--have been used to correlate experimental observations with suspected cracking mechanisms in gels. The macroscopic approach can take evaporation rate, capillary pressure, body size and many

108

other overall material characteristics into account in the model but cannot predict the CP cracking.[8-11] The microscopic theory, on the other hand, blames sudden cracking at the CP on the heterogeneity of pore sizes without, however, being able to include many obvious influence factors in a drying process such as drying rate and body size.[1] Neither of these approaches can predict all of the observed phenomena, and using both models can sometimes lead to contradictory conclusions.

The present work describes a macroscopic model that seems able to account for various observed phenomena and avoid many disadvantages of the existing theories. Some conclusions derived from the present theoretical model are surprisingly different from those previous reported but consistent with experimental observations. Although we limit our concern on gels in the present paper, the following discussions can be extended to other materials.

FRACTURE MODEL FOR TWO-PHASE MATERIAL

During the CRP and before the CP is approached, gel materials must be viewed as two-phase: they have a fluid phase and a solid skeleton and the pores are saturated by the fluid. Following conventional concepts in fracture mechanics, any catastrophic fracture is originated from preexisting fissures or (micro) cracks. Thus consider a fissure embedded in a saturated material body as shown in Fig. 1. The fissure may be partially or totally empty, or saturated so one has to distinguish between two void systems and two distinct fluid pressures for pores and fissures, respectively. The fissure fluid pressure does not necessarily equal to that of the pore and the difference between them depends on many factors: fissure configuration, permeability and fluid gradient of both fissure and pore, boundary conditions of drainage and loading of the body, and others.

Regardless of whether the pore and fissure fluid pressures are equal or not, when a remote total tensile stress is applied to the body, the fissure fluid pressure is also acted on fissure surface. If the fissure fluid pressure is in tension, it has effect of impeding the crack to propagate. In contrast, if the fissure pressure is in compression, it enhances growth of the crack. Moreover, if the fissure is partially or totally empty, the fissure pressure on the dry portion of the fissure surface equals to zero. The fissure-matrix interaction can be idealized, as shown in the left side of Fig. 1, as a crack with internal pressure in a one-phase material. The stress intensity factor K_I can be calculated in this case as

$$K_I = (T - f)\sqrt{a\pi} + (2f\sqrt{a/\pi})\arcsin(b/a) \qquad (1)$$

referring to Fig. 1, T is remote tensile stress, f fissure fluid pressure, $2a$ the total length of the crack, and $2b$ the length of the dry portion of the crack. When K_I is higher than fracture toughness of the material, catastrophic fracture occurs. If the fissure is saturated ($b = 0$), $K_I = (T - f)\sqrt{a\pi}$; if the fissure is totally empty ($b = a$),

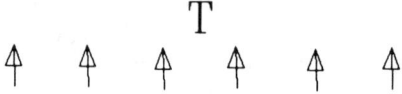

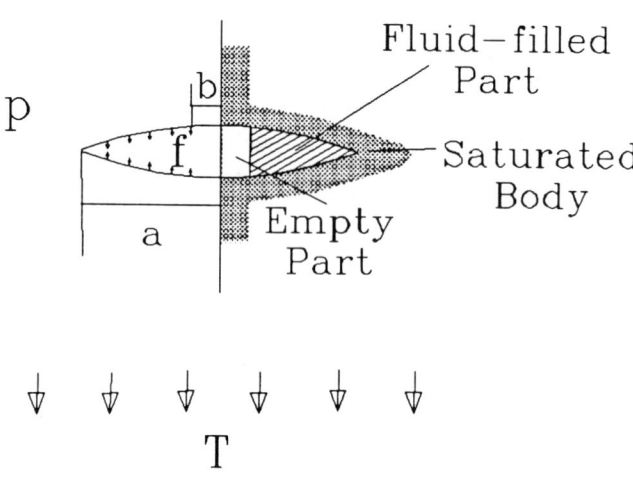

Figure 1 A crack embedded in a two-phase material (right side) and its counterpart in a one-phase material (left side). The crack has a total length $2a$ and a length of dry portion $2b$. T, f and p are remote stress, fissure pressure and pore pressure, respectively.

$K_I = T\sqrt{a\pi}$. Thus it is observed that the crack propagation is controlled by combination of total tensile stress and fissure fluid pressure rather than total stress alone. The statement pointed out herein has many practical implications in science and engineering fields and one of which is hydraulic fracture technique, widely used in petroleum engineering for secondary recovering process. Employing this method, fluid is injected under compression into a hole previously drilled in order that radial cracks around the hole grow and oil can be recovered in an efficient manner.[12]

CRP CRACKING MECHANISM

In order to explain the CRP cracking by employing the model mentioned above, it is necessary to clarify first whether there exist preexisting fissures and then how is the state of stresses in a gel body. As in any material processing procedure, during the gelation of a gel material, the solid skeleton is never formed uniformly so defects are introduced. The defects, which are expected to be distributed randomly within the bulk body, may be attributed to the inhomogeneity of concentration of solution and complicated boundary conditions of humidity and temperature. The preexisting fissures are originated from these defects.

During the CRP, both pores and fissures in gels are saturated (Fig. 2a) and it is reasonably assumed that the fissure and pore pressures are equal. According to eq. (1) for the case of b = 0, the stress intensity factor is directly related to $(T - f)$ or $(T - p)$ for $f = p$ where p denotes pore fluid pressure. The stress term $(T - p)$ is known in the mechanics literature as effective or network stress, whose calculation, as a consequence of the foregoing discussion, appears indispensable in understanding cracking mechanisms. Now consider a flat plate consisting of a purely elastic gel, whose analysis has been reported by Scherer.[8] When the evaporation proceeds in both sides of the plate, the pore pressure gradient takes place. As no free shrinkage can occur in the plate plane, residual stress is produced. The stress as well as pore pressure depends on the evaporation rate, permeability and elastic constants of material, plate thickness and drying time. Their calculation may be followed by using the Eq. 19 of ref 8 and Eq. 67 of ref 10. In the Figs. 6c and 6d of ref 8, it can be observed the spatial and temporal variation of stress and pore pressure, or T and p by using the convention adopted in the present work. In all the cases, the stress and pore pressure at the drying surface are in tension, but its difference $(T - p)$ is always in compression. Further calculation can show that the network stress $(T - p)$ does not vary spatially. Such a fact leads to the important conclusion that the network stress, which may cause the fracture, is compressive during the CRP and thus no catastrophic fracture is expected.

When the CRP is ended and the CP is approached, the menisci enter into the gel body. As the dimension of the fissures is greater than that of the pores, fissures are left empty first and then fissure pressure becomes zero (Fig. 2b). Recalling the case of $b = 0$ for totally empty fissure in the model proposed above, the stress intensity factor is related directly to the total stress. Therefore the fissures may propagate under total tensile stress T. During the subsequent FRP, fracture propagation is also controlled by the total stress.

CONCLUSIONS

Now we are ready to answer the question: Why cracks appears when the CRP stops. This is because that during the CRP, the network stress which is responsible for crack propagation in this stage is always compressive and cracking never is expected. When the CRP is ended, the fissures begin to be left empty and total tensile stress controls possible catastrophic cracking. In Fig. 2, two distinct fracture mechanisms—network stress and total stress—are illustrated. Such a conclusion could open the possibility of drying sol-gel materials before the CP, where the evaporation rate can be as high as economically feasible, without causing cracks. In order to avoid cracking after the CP, it is then necessary to accomplish control of drying rate,

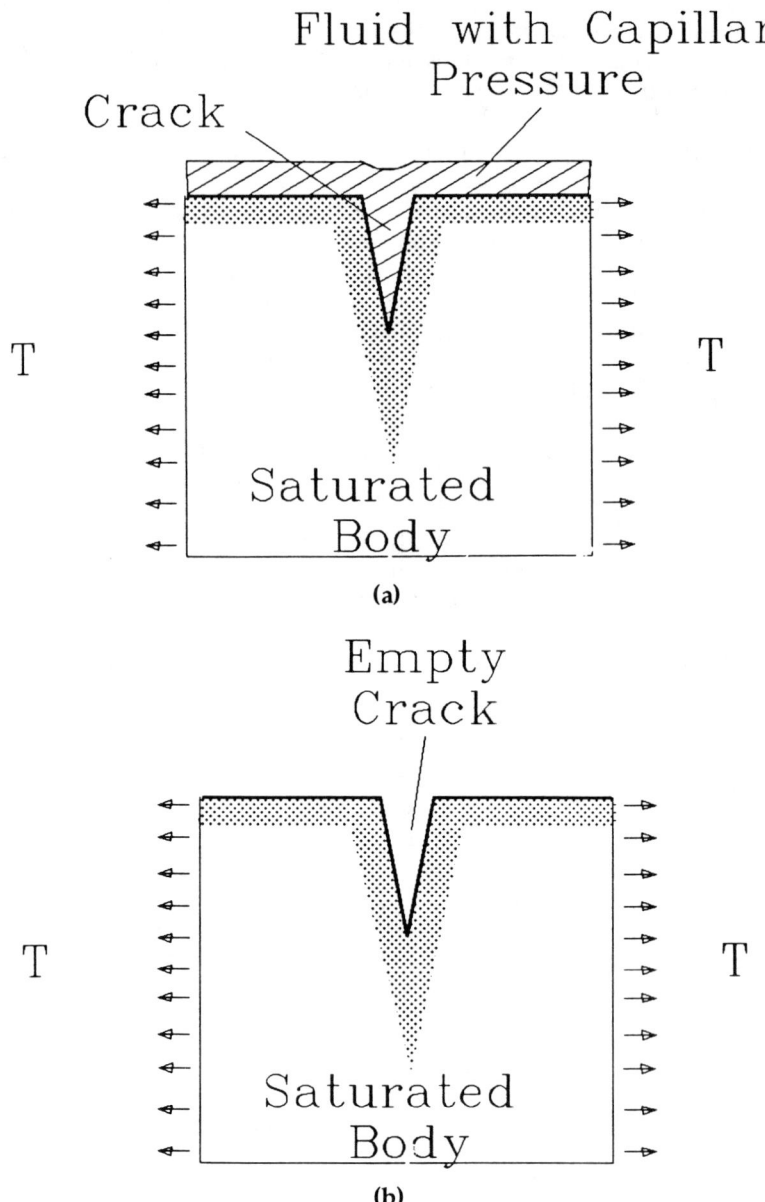

Figure 2 A crack near evaporation surface in gels. Fig. 2a illustrates a saturated fissure during the CRP and Fig. 2b, a totally empty fissure at the CP and during the FRP.

strategies for reducing capillary pressure and design of material body geometry.[11] In this way, an optimal design criterion for drying process may be established accounting for those influence factors as evaporation rate, geometric characteristics and elastic properties of gel material.

The proposed fracture model is different from the microscopic one of Zarzycki et al.[1] because this one assume that the gel is intact before the CP and fissures are formed only at and after the CP. The present model, however, considers that the fissures exist before the CP. It is interesting to show that the present theory can be also distinct from the macroscopic one of Scherer[11] in that, this one attributes the fracture propagation to the total stress rather than network stress in all drying process. Another "mixed" model[9,11] is obvious different from the present one because it simply combines the macroscopic and microscopic theories. To close the paper, it should be noted that all the theories about the fracture propagation in gel materials are established in a somewhat inferential rather than direct observation way, so experimental testings, designed to clarify important issues suggested by various theoretical models, are highly urged.

ACKNOWLEDGMENTS

The author is grateful to Dr. G.W. Scherer and Hailin Hu for many helpful discussions.

REFERENCES

1. J. Zarzycki, M. Prassas, and J. Phalippou, "Synthesis of Glasses from Gels: The Problem of Monolithic Gels," J. Mater. Sci., 17, 3371-79 (1982)
2. R.K. Dwivedi, "Drying Behavior of Alumina Gels," J. Mater. Sci. Lett., 5, 373-76 (1986)
3. R. Clasen, "Preparation and Sintering of High-Density Green Bodies to High-Purity Silica Glasses," J. Non-Cryst. Solid, 89, 335-44 (1987)
4. P. Anderson and L.C. Klein, "Shrinkage of Lithium Aluminosilicate Gels during Drying," J. Non-Cryst. Solids, 93, 415-22 (1987)
5. P.G. Simpkins, D.W. Johnson, Jr., and D.A. Fleming, "Drying Behavior of Colloidal Silica Gels," J. Am. Ceram. Soc., 72, 10 1816-21 (1989)
6. Z.P. Bazant, "Fracture in Concrete and Reinforced Concrete," in Mechanics of Geomaterials, ed. by Z.P. Bazant, John Wiley & Sons (1985)
7. J. Badillo and R. Rodriguez, "Mecanica de suelos (Soil Mechanics), Vol. 3, Third Ed., Limusa, Mexico City (1986)
8. G.W. Scherer, "Theory of Drying," J. Am. Ceram. Soc. 73 [1] 3-14 (1990)
9. C.J. Brinker and G.W. Scherer, Sol-Gel Science, Academic Press, New York (1990)
10. G.W. Scherer, "Drying Gels: VIII. Revision and Review ," J. Non-Cryst. Solids, 109, 171-82 (1987)
11. G.W. Scherer, "Stress and fracture during drying of gels," J. Non-Cryst. Solids,

121, 104-109 (1990)

12. R.P. Nordgren, "Propagation of a vertical hydraulic fracture," Soc. Petrol. Eng. J., 12, 306 (1972)

DEFORMATION DURING BINDER REMOVAL FROM MULTILAYER CERAMIC GREENWARE

Yuying Tang and Michael J. Cima
Ceramics Processing Research Laboratory, Massachusetts Institute of Technology, Cambridge, MA 02139

ABSTRACT

Deformation of green multilayer ceramic (MLC) samples during binder removal was observed. Several types of deformation were observed: thermal expansion, increased particle packing density, delamination, and healing of the delamination. A 2.4-3.7% linear shrinkage was observed, which corresponded to an increase in the particle packing density from 50% to 55% of theoretical density. Delamination occurred at both heating rates studied, 1°C/min and 10°C/min. Delamination at the lower heating rate, however, was followed by complete healing of the delamination. Only partial healing of the delamination was observed at higher heating rates. The cause of delamination was linked to the decomposition of the binder in the electrodes and possibly to the loss of plasticizer from the ceramic tape laminates.

INTRODUCTION

Binder removal from ceramic greenware is an important step in the production of high quality ceramic components. As the green ceramic component is heated, volatile species evaporate, polymers decompose, and combustion may occur in an oxidizing environment. Binder removal is often associated with the production of a variety of defects such as cracks, voids, and delaminations. An understanding of the relationships between defect formation and mass and heat transfer, changes in particle packing, and sample deformation is necessary in order to optimize processing conditions.

Previous research has focused on studying the thermolysis process and characterizing particle packing as well as sample deformation of both single layer ceramic green sheets[1,2,3] and injection molded ceramic bodies.[4-8] Little is known, however, about the thermolysis of binders from multilayer ceramic greenware such as multilayer ceramic capacitors (MLCs) and multilayer ceramic substrates.[9,10] One major

115

defect observed in MLCs is delamination, a separation of the electrode and dielectric layers. Delamination is detrimental since it results in electrode shorts. Since delamination expands the sample along one direction, it is always accompanied by sample deformation. Thus, monitoring sample deformation during the binder pyrolysis process will provide information about delamination and particle rearrangement.

This paper describes observations of sample deformation during binder burnout from MLCs. The system chosen for this investigation consisted of layered structures of green $BaTiO_3$ sheets with palladium paste electrode layers. The binder system used for ceramic tape casting consisted of a polymer, poly(vinyl butyral) (PVB), and a plasticizer, benzyl butyl phthalate (BBP). The organic ink vehicle for the electrode was ethyl hydroxyethyl cellulose (EHEC) dissolved in an aliphatic solvent. Both of these binder systems are used commercially, and the PVB-BBP system has been discussed in the open literature.[11]

EXPERIMENTAL PROCEDURE

The composition of both the ceramic tape and the printed electrodes is shown in Table 1. Tape-cast ceramic sheets consisted of 50 vol% $BaTiO_3$ particles, 47 vol% binder and plasticizer, and 3 vol% trapped gas. The green tapes were printed with electrode paste and laminated together. An MLC sample was composed of forty layers of ceramic tape, of which the middle thirty layers were printed with palladium electrode paste. The individual ceramic sheets were 34 μm thick and were separated by 3 μm-thick electrode layers. The dimensions of the sample were LxWxH = 1.62x1.32x1.34 mm^3. The thermal gravimetric analysis (TGA) curves of the dry electrode paste, the single ceramic tape, and an MLC sample are shown in Figure 1.

Table 1 Composition of Ceramic Tape and Electrode Paste.

Name	Wt%	Vol%	Description
Ceramic Tape:			
barium titanate	85	50	ceramic
polyvinyl butyl (PVB)	9	28	polymer
benzyl butyl phthalate (BBP)	6	19	plasticizer
trapped gas		3	
Electrode Paste:			
palladium	88	40	electrode
ethyl hydroxyethyl cellulose (EHEC)	12	60	binder

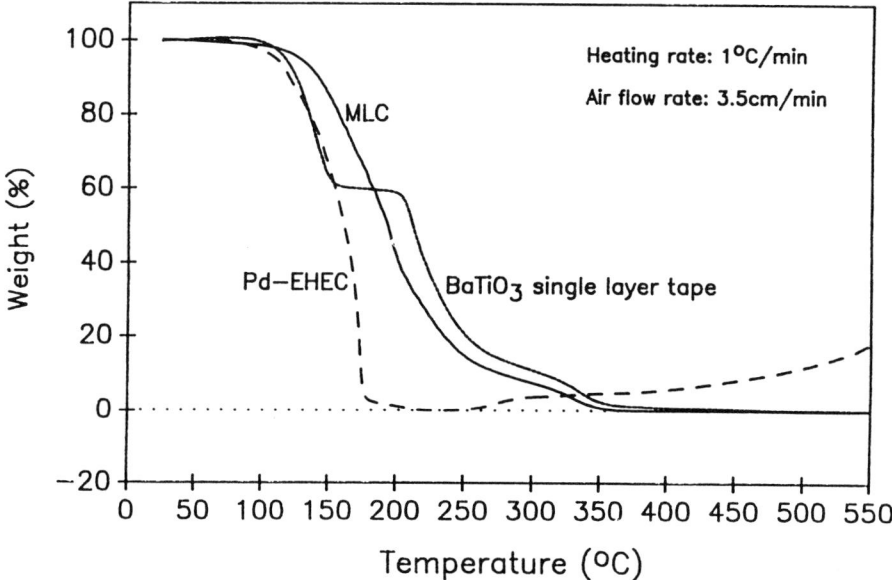

Figure 1 Thermal gravimetric analyses of electrode paste, single layer ceramic green tape, and MLC greenware.

Sample delamination produced gaps in the electrode layers of the sample, the thickness of which indicated the degree of delamination. This delamination was monitored by two methods. In the first method, a hot-stage[*]/optical microscope equipped with a TV camera was used to monitor the development of delamination *in situ*. The optical microscope was focused on the exterior surface of the MLC where the electrode termination was clearly visible. The thickness of the electrode during delamination could be measured from the TV screen at different temperatures. The second method was used to examine the interior defects of the MLC. Binder was burned out to varying degrees in a series of samples by monitoring the mass removed using TGA and quenching the sample at the desired degree of removal. The sample was then fractured through the middle, perpendicular to the electrodes, and this fracture surface was observed by scanning electron microscopy (SEM).

Three methods were chosen to measure the deformation of the entire multilayer sample during the binder removal process. In the first method, the microscope was focused on a recognizable feature on the top of a sample in a hot-stage. When the temperature was increased, the sample either expanded or contracted. The micro-

---

[*]   Linkam THM 600, United Kingdom.

117

scope focus was adjusted continually to keep the feature in focus as the sample deformed. Thus, the linear expansion or contraction during binder removal could be monitored by a simple calibration of the microscope. The approximate error in calibration was ±2 μm, which is about ±0.15% of linear deformation. The advantage of this method is that no mechanical stress is applied to the sample while its dimensions are measured. Unfortunately, only one point, or at most several points, on the surface can be monitored during the heating process, and the sample cannot be turned around in order to observe changes in other dimensions. In the second method, TGA was used to monitor the weight loss of the sample, which was then quenched when the desired amount of binder was removed. Different samples were burned out to different stages, followed by quenching. The size of the sample could then be measured using an optical microscope before and after the binder was removed. In the third method, TGA was used again to heat one sample until 5% of the binder was removed, followed by quenching. The dimensions of the sample were measured. The same sample was then reheated at the same heating rate to a higher temperature to remove more binder, quenched, and the dimensions were again recorded. This process was repeated until all of the binder was removed.

OBSERVATIONS

Delamination observation via Method 1: When a sample was heated at 10°C/min, delamination was first observed on the surface of the sample at 180°C. This delamination continued to grow until the temperature reached 230°C, after which it started to shrink or heal. The thickness of one delaminated print layer versus temperature during the heating process is shown in Figure 2. Residual delamination of approximately two-thirds the maximum separation remained at 500°C after all of the binder had been removed. Since the delamination always started from inside the sample (usually from the sample center), it was expected that the delamination in the middle of the sample would be more severe and would start at lower temperature than that observed on the surface. The residual delamination in the center of this sample, as measured from an SEM micrograph of a fracture surface through the center of the sample, was 54 μm. This residual delamination was more than six times larger than that observed on the surface, which was 8.8 μm.

Delamination observation via Method 2: For a series of samples all heated at 1°C/min to a number of final temperatures, no delamination could be observed from the surface of the samples. The SEM micrographs of fracture surfaces of the samples, however, revealed the same delamination-and-healing process observed at 10°C/min with Method 1. The beginning of delamination was observed in the sample heated to 155°C, which corresponds to 20% binder removal. Delamination was severe (10 μm) in the sample heated to 165°C, which had lost 30% of the binder. High temperature caused the extent of delamination to diminish such that the samples were completely healed at 190°C, or 50% binder removal.

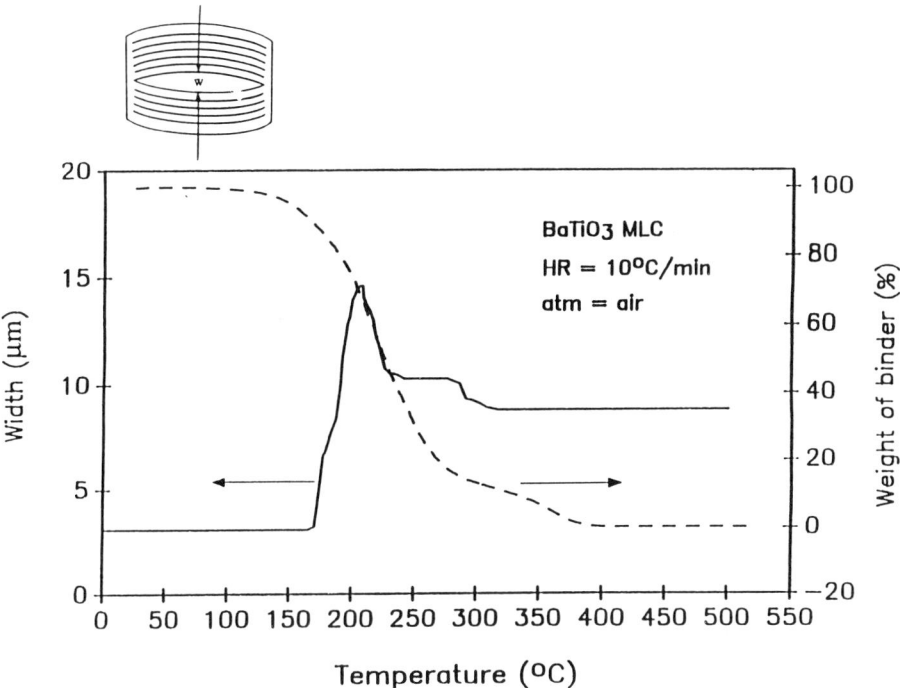

Figure 2 Thickness of one delaminated print layer vs. temperature.

Sample deformation measurement via Method 1: Figure 3 shows the deformation perpendicular to the dielectric-electrode interface direction (L) versus temperature at different heating rates, as measured with an optical microscope. Samples heated at all rates experienced expansion below 160°C, and the sample heated at 10°C/min experienced a very large expansion above 160°C followed by a sharp shrinkage. At lower heating rates, the expansion was smaller and the net shrinkage was larger. A slow healing process was observed at all three heating rates. Figure 4 shows the percent linear expansion versus temperature in different directions for a sample heated at 10°C/min. After a slow expansion before 160°C, a rapid expansion was observed in all directions followed by a sharp shrinkage which started at 210°C. The expansion in the direction perpendicular to the dielectric-electrode interface (L) was greater than that in the parallel directions (M and P). The magnitude of net shrinkage in the three directions, l, was $l_M > l_P > l_L$. The percent linear expansion versus temperature in the M and P directions was also observed for samples heated at 1°C/min. The net shrinkages were the same as their respective values at 10°C/min. A sharp shrinkage occurred at all heating rates and along all directions. At a heating rate of 1°C/min, the linear net shrinkages in the L, M, and P directions were

119

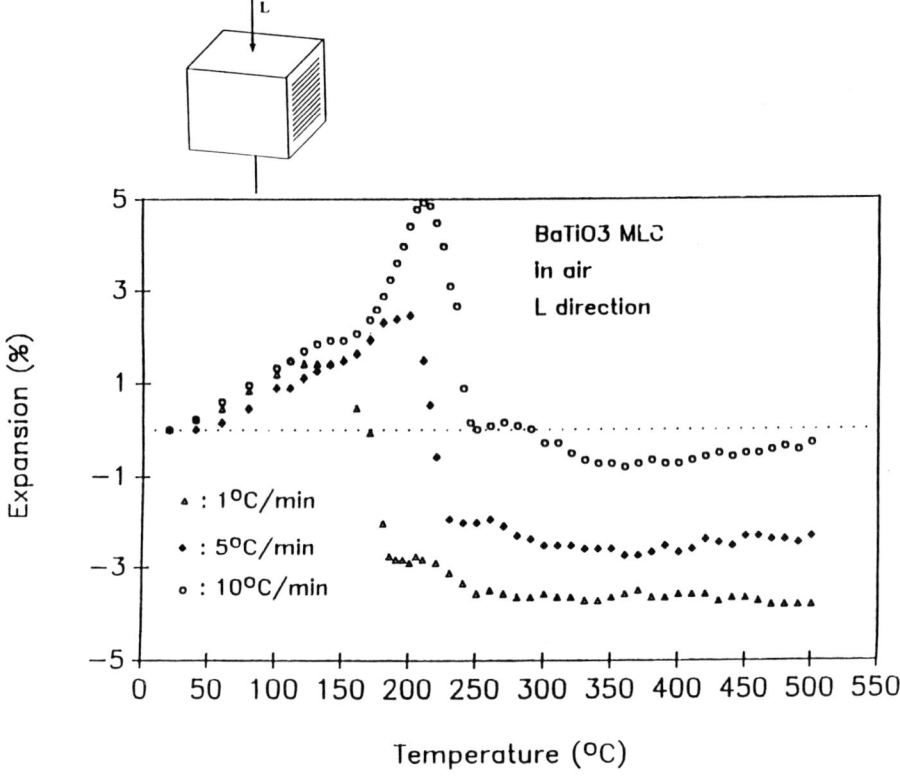

Figure 3 Dimensional change of an MLC sample in the L direction vs. temperature at different heating rates.

3.7%, 3.5%, and 2.4%, respectively. This corresponds to a 9% volume shrinkage.

Sample deformation measurement via Methods 2 and 3: Figure 5 shows the volume shrinkage of samples quenched at different stages of binder removal. The solid line represents a series of samples which were burned out to different stages, followed by quenching. The dashed line represents one sample which was heated in cycles until all the binder was removed. Table 2 lists the quenching temperatures of the samples used in the two experiments above.

In the early stages of binder removal, the decrease in sample volume was very close to the volume of binder lost by evaporation. When 20% binder was removed, this shrinkage stopped. A slight expansion due to delamination, followed by a slow healing process, occurred in samples when more binder was removed.

120

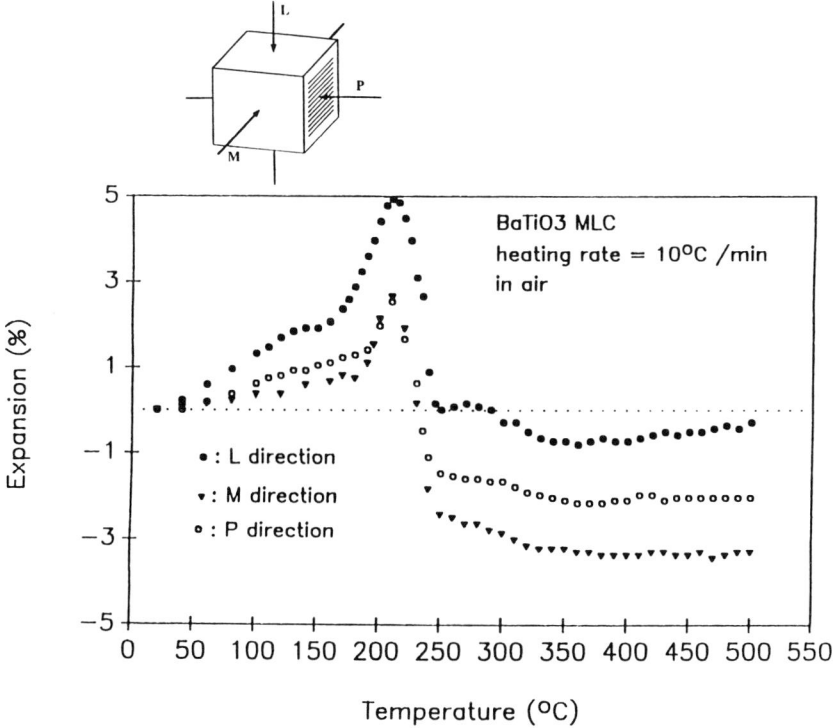

Figure 4 Dimensional change of an MLC sample in different directions vs. temperature at a heating rate of 10°C/min.

Table 2 The Quenching Temperature at Varying Degrees of Binder Removal.

Amount of binder removal (wt%)	5%	10%	20%	30%	40%	50%	60%
T(°C): (series of samples used)	123	138	155	165	179	189	198
T(°C): (one sample used)	123	133	140	151	171	184	195

The linear shrinkage of the cyclically heated sample at different stages of burnout was also observed. The linear shrinkage along the L direction reached a maximum at 20% binder removal. The resulting linear shrinkages in the L, M, and P directions due to binder removal were 3.5%, 3.1%, and 2.5%, respectively.

121

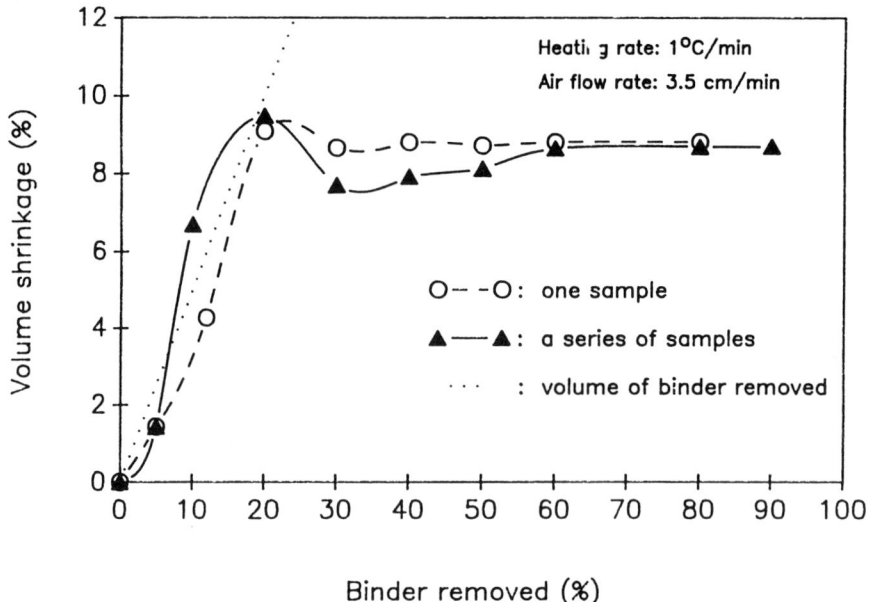

Figure 5 Volume shrinkage of MLC samples vs. amount of binder removed.

DISCUSSION

The three types of experiments described above provide insight into the origin of the deformation observed. Four types of deformation were observed at low heating rates, such as 1°C/min. During the first stage (below 120°C), thermal expansion and stress relaxation were the main causes of sample deformation. A typical thermal expansion coefficient of plasticized PVB binder is 2×10^{-4}/°C.[12] Thus, a 1% linear expansion is expected between room temperature and 120°C for an MLC sample consisting of 50 vol% polymer, which is consistent with the expansion shown in Figures 3 and 4. For a green MLC sample, both the casting process and the lamination process caused compressive stresses in the L direction. It was not surprising to observe a larger linear expansion in the L direction than in the M and P directions, which are 1.5% and 0.5% at 120°C, respectively.

The second stage of deformation was a rapid shrinkage of the sample which occurred between 120°C and 155°C. During this period, 50% of the BBP plasticizer evaporated, causing a reduction by 20% in the total amount of binder present. The decrease in volume of the sample was equal to the volume of binder lost by evaporation. The particle packing was clearly not static during binder removal. As stated above, the initial particle packing density was as low as 50 vol%, but grew to 55 vol% following binder removal. Substantial particle rearrangement is consistent with the capillary flow known to occur as PVB-based binders are removed.[13] Capillary forces created by the retreating vapor interface of the molten binder within the pores could rearrange the particles to higher packing density.

Delamination occurred from 155°C to 180°C, or 20% to 40% binder removal. At faster heating rates, delamination would occur during the period at which the ceramic matrix was shrinking. This temperature range coincides with the removal of all the electrode and half the plasticizer, as shown in Figure 1. The porosity of the sample remained low due to repacking of the particles. Conservation of volume is verified by the data shown in Figure 5. The gas generation rate of the ink vehicle must, therefore, be compensated for by flowing through the thin electrode toward the electrode termination. The obvious flow resistance must also be accompanied by a pressure increase within the electrode. At high enough gas generation rates, this pressure is apparently enough to delaminate the sample.

Delamination was observed to be a rather common event even at low heating rates. In all cases, however, considerable healing of the delamination occurred as the temperature was raised. Small delaminations produced at low heating rates appeared to heal completely. Thus, delamination may be a common occurrence in the processing of MLCs but is not detected in the fired component because the delamination does not occupy any volume after heating to high temperature.

CONCLUSIONS

Sample deformation was observed in order to monitor changes in particle packing and formation of defects during binder removal. MLC greenware experienced four types of deformation when heated in air. At low temperatures, only thermal expansion and stress relaxation were observed. Particle packing density increased at higher temperature as the binder was removed due to the capillary action of the molten binder. Delamination occurred even at low heating rates, but healed upon further heat treatment. High heating rates produced such large delamination strains that complete healing did not occur. Delamination was associated with the rapid decomposition of the electrode binder causing increased gas pressure between the laminates.

REFERENCES

1. M.J. Cima, J.A. Lewis, and A.D. Devoe, "Binder Distribution in Ceramic Greenware During Thermolysis," *J. Am. Ceram. Soc.*, **72**, 1192-99 (1989).
2. D.W. Sproson and G.L. Messing, "Organic Removal Processes in Closed Pore Powder-Binder System"; pp. 528-37 in Ceramic Transactions: Ceramic Powder Processing Science. Edited by G.L. Messing, E.R. Fuller, Jr., and H. Hausner. American Ceramic Society, Westerville, OH, 1988.
3. Y.-N. Sun, M.D. Sacks, and J.W. Williams, "Pyrolysis Behavior of Acrylic Polymers and Acrylic Polymer/Ceramic Mixtures"; pp. 538-48 in Ceramic Transac-

tions: Ceramic Powder Processing Science. Edited by G.L. Messing, E.R. Fuller, Jr., and H. Hausner. American Ceramic Society, Westerville, OH, 1988.

4. J.K. Wright, M.J. Edirisinghe, J.G. Zhang, and J.R.G. Evans, "Particle Packing in Ceramic Injection Molding," *J. Am. Ceram. Soc.*, **73**, 2653-58 (1990).

5. C.A. Sundback, M.A. Costantini, and W.H. Robbins, "Part Distortion During Binder Removal"; pp. 191-200 in Ceramic Materials and Components for Engines. Edited by V.J. Tennery. American Ceramic Society, Westerville, OH, 1989.

6. J.K. Wright, J.R.G. Evans, and M.J. Edirisinghe, "Degradation of Polyolefin Blends Used for Ceramic Injection Molding," *J. Am. Ceram. Soc.*, **72**, 1822-28 (1989).

7. J. Woodthorpe, M.J. Edirisinghe, and J.R.G. Evans, "Properties of Ceramic Injection Molding Formulation III. Polymer Removal," *J. Mater. Sci.*, **24**, 1028-49 (1989).

8. B.C. Mutsuddy, "Oxidative Removal of Organic Binders from Injection-molded Ceramics"; pp. 397-408 in Proceedings of the International Conference on Non-Oxide Technical and Engineering Ceramics, National Institute of Higher Education, Limerick, Ireland, 1985.

9. R.A. Gardner and R.W. Nufer, "Properties of Multilayer Ceramic Green Sheets," *Solid State Technol.*, **17** [5] 38-43 (1974).

10. J.G. Pepin, "Electrode-Based Courses of Delaminations in Multilayer Ceramic Capacitors," *J. Am. Ceram. Soc.*, **72**, 2287-91 (1989).

11. R.E. Mistler, D.J. Shanefield, and R.B. Runk, "Tape Casting of Ceramics"; pp. 411-48 in Ceramic Processing Before Firing. Edited by G.Y. Onoda and L.L. Hench. Wiley, New York, 1978.

12. Information provided by Monsanto Corp., Amherst, MA.

13. M.J. Cima, M. Dudziak, and J.A. Lewis, "Observation of Poly[Vinyl Butyral]-Dibutyl Phthalate Binder Capillary Migration," *J. Am. Ceram. Soc.*, **72**, 1087-90 (1989).

SOLID OXIDE FUEL CELL CERAMICS THROUGH COLLOIDAL PROCESSING

K. Kendall
ICI Advanced Materials, P.O. Box 11, Runcorn Cheshire UK

ABSTRACT

The performance and economics of Solid Oxide Fuel Cells (SOFC) depend substantially on the ceramic components, especially the zirconia membrane which conducts oxygen ions from cathode to anode. SOFC membranes made from spray dried yttria stabilized zirconia were brittle and exhibited a bending strength which was governed by process related flaws, particularly agglomerates in the original powder. Direct measurement of agglomerate strength using a nanoindenter showed that a considerable force was required to break down the powder agglomerates. This agglomerate strength varied widely through a single spray drier batch. Varying the spray drier conditions allowed weaker agglomerates to be formed, and these gave stronger, more reliable membrane materials.

INTRODUCTION

Solid Oxide Fuel Cells (SOFC) [1-3] operate around 900°C where zirconia can conduct oxygen ions with resistivity near 0.1 Ωm. Under this condition, if oxygen is passed along one surface of the zirconia, and hydrogen down the other, then oxygen ions are transported through the zirconia, to react with the hydrogen, forming water. A gradient of oxygen is thus maintained across the membrane, and is equivalent to a potential difference close to 1 volt between electrodes placed on the surfaces. A current between 0.1 and 0.5 amp can be drawn from each square centimeter of membrane surface under these circumstances.

A stack of such zirconia membranes, coated with suitable electrodes, and properly manifolded to convey fuel and oxidant to the surfaces, could be used to generate electrical power. This device shows substantial benefits over internal combustion engines in terms of efficiency, which may be twice that of a diesel, and in terms of effluent, which contains no NOx. In particular, SOFC is perceived as beneficial in small scale combined heat and power applications where low maintenance cost is critical. The Commission of the European Community is now funding several programmes to investigate SOFC. ICI is collaborating with ABB and British Gas to develop zirconia for this purpose.

The major problem with this SOFC device is the integrity of the thin ceramic membrane when made by powder methods. If the zirconia is to be sufficiently thin, around 100 μm, leak tight and thermal shock resistant, then it must generally be made by a fluid, that is a colloidal, processing route. It must also exhibit high strength and can therefore contain no large flaws. The purpose of this paper is to study processing flaws found in ceramic membranes made by plastic processing spray dried zirconia powders. The individual spray dried agglomerates were strength tested using a nanoindenter. Changes in agglomerate strength due to spray drier parameters were monitored.

EXPERIMENTAL

Spray dried zirconia powder (SY ULTRA, Z Tech Pty) was mixed with a 50% aqueous solution of polyvinyl alcohol using a twin-roll mill. The final volume fraction of the plastic mix was 0.51. After degassing, the material was ram extruded to form rods 1 mm in diameter which were dried, burned out to 500°C and sintered for 1 hour at 1400°C. Three point bend testing allowed the strength to be measured and plotted in Fig. 1. The mean strength of the zirconia was 1070 MPa and the

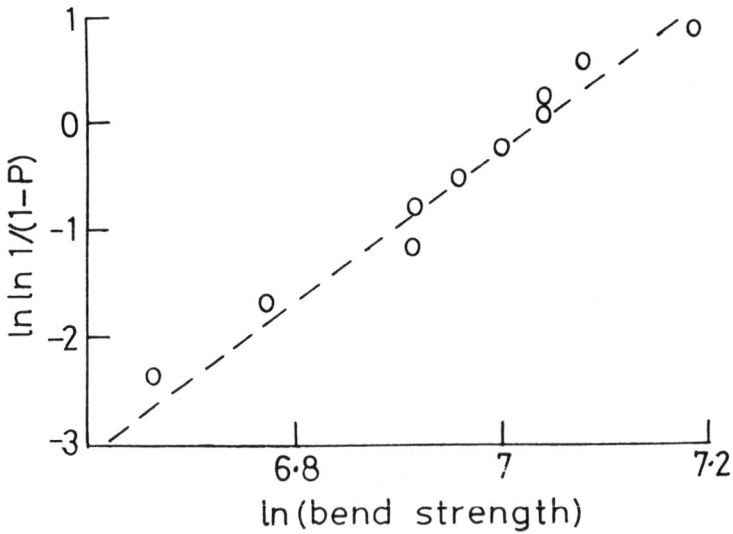

Figure 1 Strength of zirconia ceramic made by plastic processing of spray dried zirconia powder.

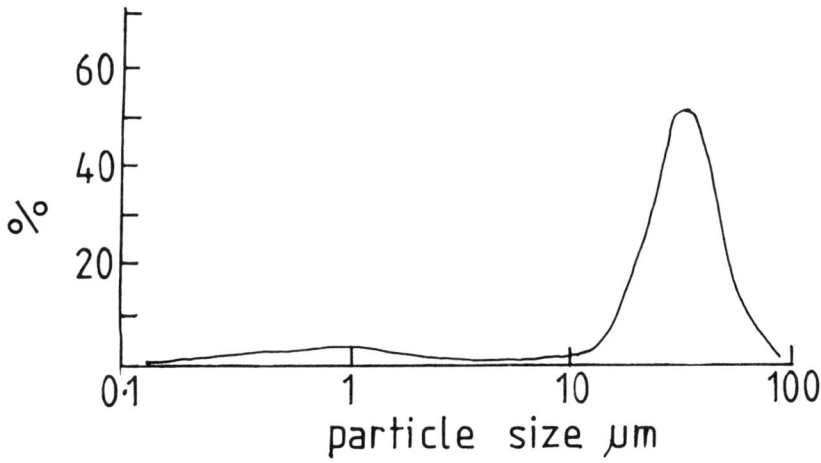

Figure 2 Malvern size analysis of spray dried agglomerates before and after one minute ultrasonic treatment in water.

Weibull modulus was 7. It was suspected that the low Weibull modulus was a result of agglomerates resisting breakdown during the plastic processing. Particle size analysis (Fig. 2) of the spray dried powder using the Malvern Mastersizer laser light scattering instrument revealed that the agglomerates did not break down after one minute treatment in the 50 watt ultrasonic bath, suggesting that the agglomerates were strongly cohesive. After twin-roll milling of the powder with the polymer, size analysis gave results very close to those in Fig. 2, showing the agglomerate persistence through mixing.

Agglomerate Strength Tests

In order to confirm the idea that the spray dried agglomerates were resistant to breakdown, individual agglomerates were compression tested using a modified nanoindenter apparatus [4-6]. A single, spherical agglomerate between 10 and 70 μm in diameter was placed on a smooth zirconia substrate and a flat zirconia plate was brought down to compress the agglomerate while measuring the applied load and displacement. The loading was carried out using an electrical actuator which incremented the force in 0.25 μN steps. Meanwhile the displacement was measured using a three plate capacitance gauge giving a 0.3 nm resolution.

Typical load displacement traces are shown in Fig. 3. The upper curve corresponds to a strong agglomerate which was made from a spray dried suspension 0.21 μm in mean particle size. There was no binder in this suspension, so the agglomerates

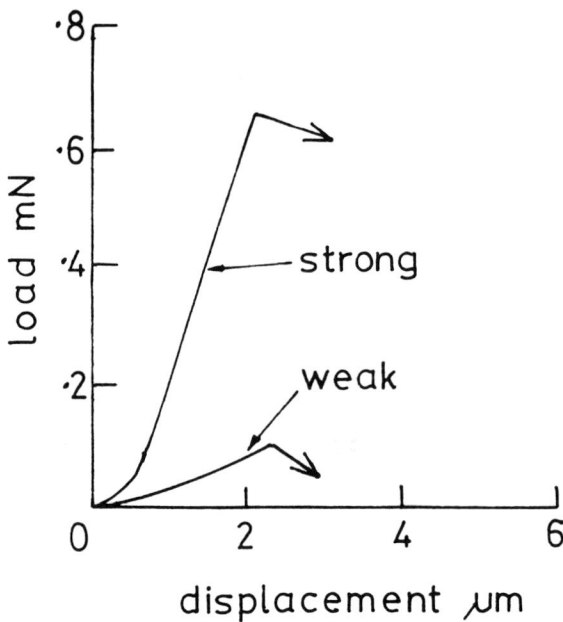

Figure 3 Nanoindenter results for compression of agglomerates made from two spray dried zirconia dispersions.

were held together by van der Waals attractions only. The curve was curved at low loads, but became straighter as the load increased. This linear part of the curve was characteristic of yielding within the agglomerate, probably by particle sliding and reordering. At a certain load, the agglomerate fractured, as shown by the jump in the load displacement line. The tensile fracture strength σ^* of the agglomerate was calculated from the equation [7]

$$\sigma^* = 2.8 \, F/\pi D^2 \tag{1}$$

where F was the peak load and D was the agglomerate diameter. The strength was found to be 0.59 MPa for this 31 μm diameter agglomerate.

To make weaker agglomerates, the spray drier conditions were altered. Instead of grinding the calcined zirconia to 0.21 μm, the comminution was stopped at 0.6 μm and the dispersion spray dried as before. Testing an agglomerate from this sample gave the lower nanoindenter curve in Fig. 3, indicating a strength six times lower than before. Such a weakening of the agglomerate would be expected from a model of spherical particles held together by van der Waals forces at elastic contact spots

[8,9]. That theory showed the strength of a perfect agglomerate to be given by

$$\sigma^* = 15.6 \, \varphi^4 \, \Gamma/d \qquad (2)$$

where φ is the volume packing of solids in the agglomerate, Γ is the work of adhesion between the particle surfaces and d the particle diameter. Clearly, the particle diameter d should exert a considerable influence on agglomerate strength. It was observed that the agglomerate volume packing, measured by mercury porosimetry, also fell when comminution was curtailed.

Less expected was the wide variation in agglomerate strengths measured on the nanoindenter. Fig. 4 shows a series of load displacement curves obtained from the same batch of spray dried powder. There was significant variation in the curves, both in the yielding of the agglomerates, and in the ultimate fracture load. The Weibull modulus of the agglomerate fracture strength from these tests was 1.5, suggesting a great variation in agglomerate structure from the spray drier. This result suggests that spray drying could be much improved to provide more uniform granules which could then be compacted into more uniform ceramic mouldings.

The weaker agglomerate spray dried powder was twin-roll mixed with polymer solution, deaired and extruded as before to make 1 mm diameter rods which were dried, burned out and sintered prior to bend testing. The results shown in Fig. 5

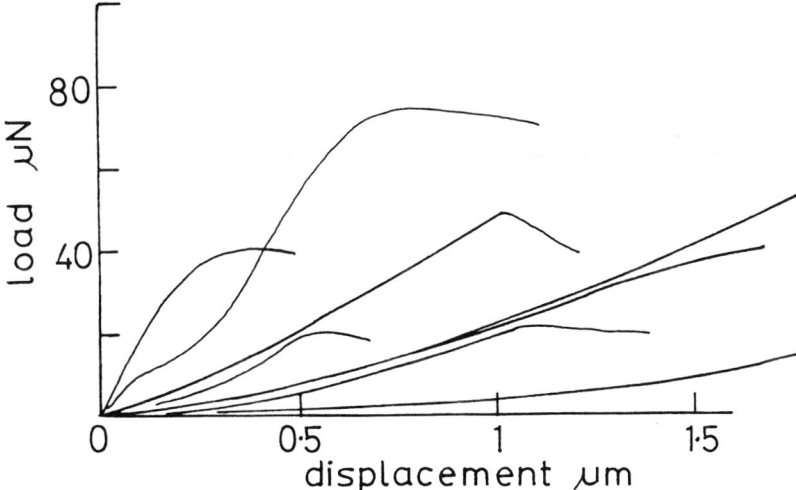

Figure 4 Nanoindenter load displacement curves for a series of spray dried agglomerates from the same batch, showing the variation in strength.

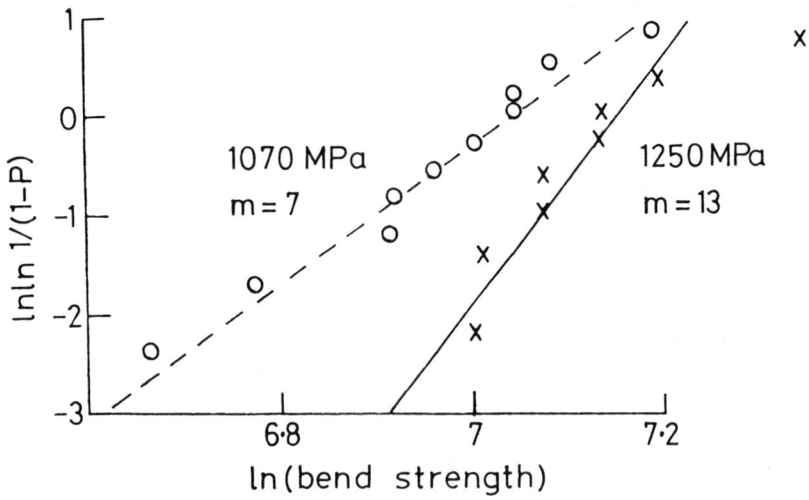

Figure 5 Improvement in strength and Weibull modulus of zirconia rods due to weakening of the spray dried agglomerates.

indicate that the final product improved significantly as a result of using weaker agglomerates in the mixing process. The mean strength rose to 1250 MPa while the Weibull modulus almost doubled to 13. It was evident that the weaker agglomerates were now being broken down during plastic mixing, reducing the number of large agglomerate defects in the product and preventing low strength samples in the test batch. The benefit of this improvement was noticeable when coating fuel cell electrodes onto the membranes. A lower failure rate was observed for the stronger, more reliable batch.

CONCLUSIONS

SOFC membranes can be made from zirconia powders by colloidal processing. When spray dried zirconia powder is used, the strength and reliability of the fired product may be low because the large agglomerates are not broken down during plastic mixing and remain as large defects in the final product. Direct measurement of agglomerate strength by nanoindenter testing has demonstrated that spray dried agglomerates held together only by van der Waals forces may exhibit high and variable strength. Nanoindenter study also shows that agglomerates may be made weaker by increasing the size of particles within the agglomerate, as expected theoretically. Such weaker agglomerates give stronger, more reliable sintered zirconia products because better agglomerate breakdown is achieved during mixing.

ACKNOWLEDGMENTS

Much credit for this work belongs to Dr. T.P. Weihs who carried out the nanoindenter measurements at Oxford University.

REFERENCES

1. B. Riley, Solid oxide fuel cells - the next stage, J. Power Sources 29 (1990) 223-238.
2. R.F. Singer, F.J. Rohr and A. Belzner, Solid oxide fuel cells:CFP design and cell performance, 1990 Fuel Cell Seminar, Phoenix Arizona (Courtesy Associates Washington DC 1990) 111-114.
3. A.J. Appleby, From Sir William Grove to today: Fuel cells and the future, J. Power Sources 29 (1990)3-11.
4. K. Kendall and T.P. Weihs, Adhesion of nanoparticles within spray dried agglomerates, J. Phys D: Appl Phys (1991) in press.
5. T.P. Weihs, S. Hong, J.C. Bravman and W.D. Nix, Mechanical deflection of cantilever microbeams; a new technique for testing the mechanical properties of thin films, J. Mater Res. 3 (1988) 931-942.
6. J.B. Pethica, R. Hutchings and W.C. Oliver, Hardness measurements at penetration depths as low as 20 nm, Phil Mag A48 (1983) 593-606.
7. Y. Hiramatsu and Y. Oka, Determination of the tensile strength of rock by a compression test of an irregular test piece, Int. J. Rock Mech Min Sci., 3 (1966) 89-99.
8. K. Kendall, N. McN Alford and J.D. Birchall, Elasticity of particle assemblies as a measure of the surface energy of solids, Proc. R Soc. Lond A412 (1987) 269-283.
9. K. Kendall, Behaviour of particle assemblies - relevance to ceramic processing, Materials Forum 11 (1988) 61-70.

NUMERICAL SIMULATION OF THE EXTRUSION OF PLASTIC BODIES

William B. Carlson, Jingmin Zheng, and James S. Reed
New York State College of Ceramics at Alfred University, Alfred, NY 14802

ABSTRACT

The basic flow behavior of extruded electrical porcelain pastes were numerically simulated via a finite difference method (FDM). Isothermal and incompressible fluid assumptions were used to approximately model the extrusion of the pastes in a square entry sheet die. It was assumed that the flow in the die-entrance region was of a viscous nature and in the die-land region obeys plug flow assumptions. The viscosity relationship used in the rheological model is of the power law form. Stream function and vorticity transport equations were used in the FDM computations. Only flow at steady-state is modeled. Calculation of the velocities at steady-state, is based upon the two step vorticity stream function procedure. Viscosity was recalculated according to power law during each iteration and applied to the vorticity calculation.

INTRODUCTION

Extrusion has been successfully employed to shape ceramic objects for a very long time. Typical ceramic extrusions include traditional products such as bricks and tiles to more advanced products such as honeycomb catalysts and silicon nitride for use in the automobile industry. A rapidly growing area of business lies in the processing of ceramics for the electrical power and the micro-electronics industries. The electrical insulator manufacturers are trying to control the quality of their products by improving their general understanding of the extrusion process.

When forming ceramics by paste extrusion, several factors need to be taken into account.[1] The two most important are paste rheology and die shape. These two parameters affect not only the required extrusion pressure but also the extrusion velocity and flow patterns. Results based on laboratory extrusion tests show that both the pressure loss in a converging die-entrance region and in the die-land region depend on the extrusion rate, and the paste viscosity.[1-3] The size of the static zone affects the quality of the product and the die design. Flow-visualization studies in

132

the die entrance region[4] have proven important to the understanding of fluid properties in other materials.

Previous computer simulations via the finite element method (FEM) have been used to calculate the streamlines of extruded plastic bodies in a converging entrance die.[5,6] The finite difference method (FDM) has also been used as a powerful tool to simulate other fluid flow problems.[7] In this paper, stream function and vorticity transport equations were used in the FDM computations along with a power law viscosity model. To calculate the vorticity at steady-state, the viscosity was first set to be a constant. The viscosity was re-calculated according to a power law before the next calculation of vorticity until steady state convergence was reached. The dependence of the streamlines on the exponent in the power law was studied and qualitatively compared to the flow pattern expected in extrusions of electrical porcelains.

FORMULATION

Velocity Field

The simplest type of convergent flow to model is the two-dimensional sheet flow problem. Consider a slit of uniform thickness H_O before contraction and H after contraction. This is the case when thin substrates and tiles are formed by extrusion. In such a case, we may neglect the out of plane velocity, therefore, the velocities and shear rates become:

$$\vec{v} = v_x(y)\,\vec{e_x} + v_y(x)\,\vec{e_y} \tag{1}$$

and

$$\gamma_{xy} = \frac{\partial v_x}{\partial y} \qquad \gamma_{yx} = \frac{\partial v_y}{\partial x} \tag{2}$$

where v is flow velocity and γ, is shear rate.

Rate Changes of Mass and Motion

The flows are assumed to be isothermal, incompressible, and slow. Gravitation is neglected so that the governing equations for motion, the continuity and the Navier-Stokes equations, take the form:

$$\nabla \cdot \vec{v} = 0 \tag{3}$$

$$-\nabla p + \mu \nabla^2 \vec{v} = 0 \tag{4}$$

where v is the velocity field, p is pressure and μ is viscosity.[4]

For the case of the two-dimensional flow, equations 3 and 4 become:

$$0 = \frac{\partial v_x}{\partial x} + \frac{\partial v_y}{\partial y} \tag{5}$$

and

$$0 = -\frac{\partial p}{\partial x} + \mu\left(\frac{\partial^2 v_x}{\partial x^2} + \frac{\partial^2 v_x}{\partial y^2}\right) \tag{6}$$

$$0 = -\frac{\partial p}{\partial y} + \mu\left(\frac{\partial^2 v_y}{\partial x^2} + \frac{\partial^2 v_y}{\partial y^2}\right) \tag{7}$$

Vorticity Equations

The pressure is eliminated from equations 6 and 7 by cross-differentiating equation 6 with respect to y and equation 7 with respect to x. Subtracting one from the other yields:

$$0 = \mu\left(\frac{\partial^2 \xi}{\partial x^2} + \frac{\partial^2 \xi}{\partial y^2}\right) = \mu \nabla^2 \xi \tag{8}$$

where

$$\xi = \frac{\partial v_x}{\partial y} - \frac{\partial v_y}{\partial x} \tag{9}$$

is defined as vorticity. Equation 8 is the steady-state vorticity transport equation for slow incompressible flow. Notice that from equation 4 through equation 8, the viscosity is presumed to be constant as for Newtonian fluids. For a variation of viscosity with shear rate, as in the non-Newtonian fluids, the vorticity transport equation has a viscosity function and the variation of viscosity requires that $0 = \nabla^2 \mu \xi$ for a finite numerical approach.

Stream Functions

A stream function ψ is defined by

$$\frac{\partial \psi}{\partial y} = v_{x'} \qquad \frac{\partial \psi}{\partial x} = -v_y. \qquad (10)$$

The function $\psi(x,y)$ represents the streamlines of flow.

Combining the vorticity equation and the stream function equation 10 gives an elliptic Poisson equation:

$$\nabla^2 \psi = \xi \qquad (11)$$

Viscosity

Viscosity is defined as:

$$\mu = \frac{\tau}{\gamma} \qquad (12)$$

In this paper the Herschel-Bulkley equation is used to describe the viscosity of a plastic fluid:

$$\tau = Y + K\gamma^n \qquad (13)$$

Where τ is the paste shear stress, Y is the yield stress, and K and n are material dependent constants. If velocity rate is very low, a static zone is assumed to develop. If Y is 0 and n equals 1, we have Newtonian flow. In the Bingham fluids Y=Constant and n=1. For fluids with $0<n<1$ we have a pseudoplastic response.[8]

Simulation Procedure

A central finite difference scheme was used to approximate the differential vorticity and stream function equations respectively. A standard five node pattern with $0(h^2)$ error was used in each function, and the analysis was run for a number of cycles until steady-state conditions were observed. This iteration scheme is shown in Figure 1. Boundary conditions were assumed from the continuity of velocity in the control volume (see Fig. 2). As the power law exponent was varied in each simulation, the material shear and, therefore, the static zone at the die entrance was altered.

RESULTS AND DISCUSSION

In the simulations, the shear consistency index K is held constant. This parameter is difficult to assess for electrical porcelains because of experimental difficulties, therefore, an average K was assumed to be closely approximated as 2 kPa(s)^n where n is the shear thinning exponent. This exponent was taken from experimental

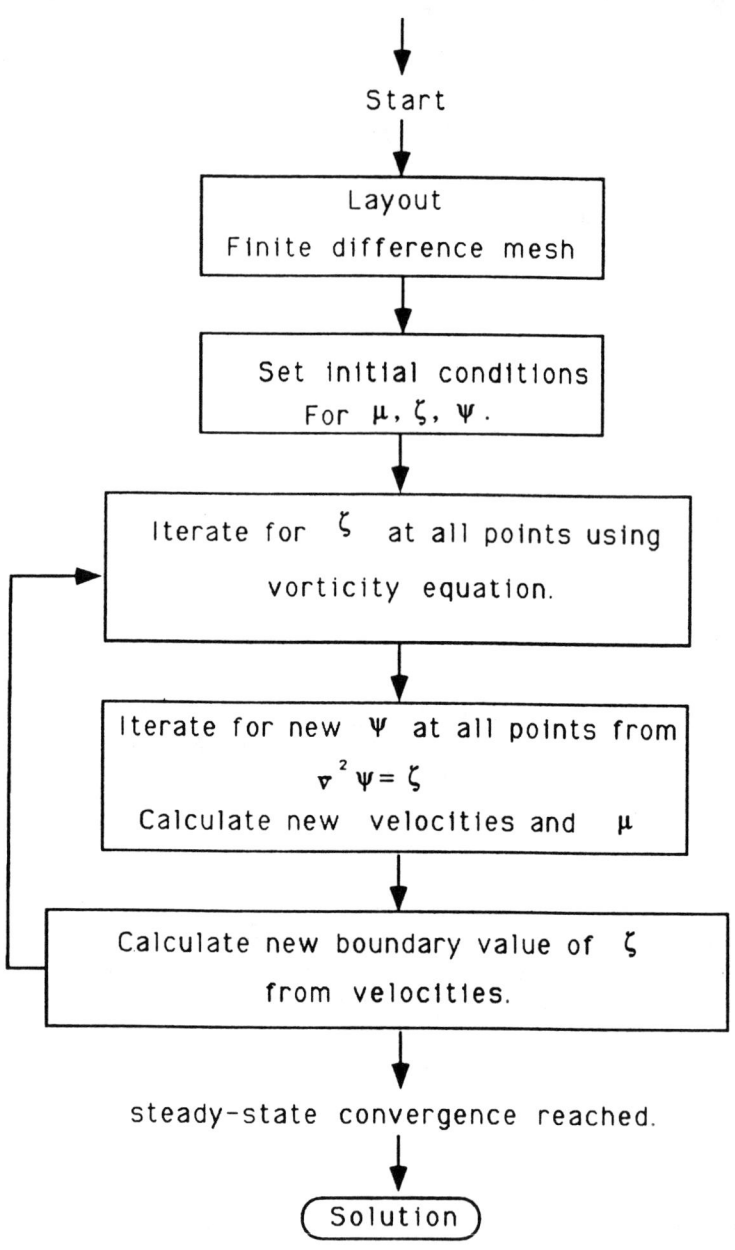

Figure 1 The finite difference iteration scheme for the calculation of the vorticity-stream functions.

136

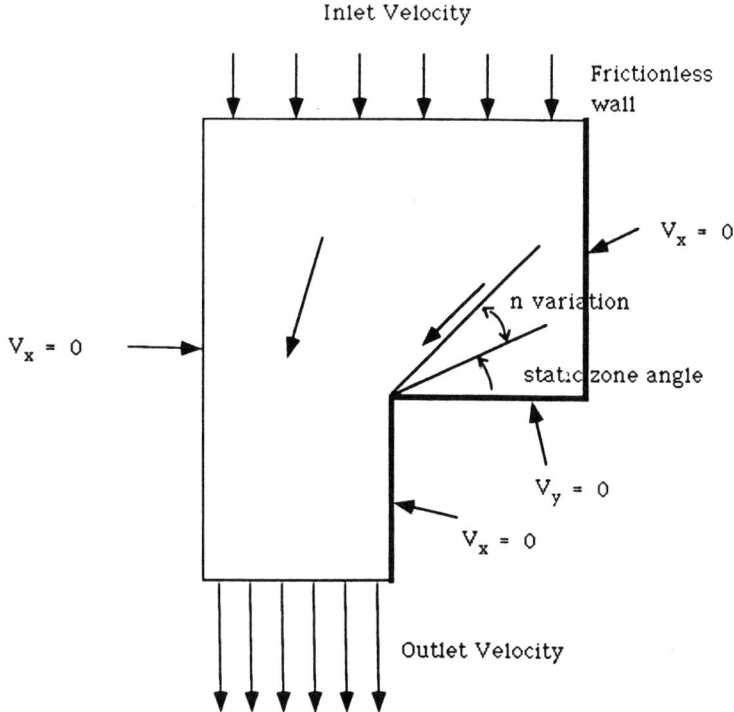

Figure 2 Velocity boundary conditions.

extrusion results[9] for electrical porcelain and varied in the numerical simulations between 0.25 and 0.50. Newtonian fluid simulations were run as baseline cases to compare with the pseudoplastic simulations.

A Newtonian streamline result is presented in Figure 3. The streamline velocities are higher in the entry zone with a relatively small static zone.

In Figures 4 and 5 the shear thinning exponent, n, is 0.4 and 0.25 respectively. Decreasing the exponent between these values develops a larger static zone. Flow studies of polymer materials show that the vortex region in the square-die entry also changes with rheological properties. Investigators[10] have found that the strain-thickening elongational viscosity is primarily responsible for a vortex growth in the entry region. Results in this paper have thus far shown that shear-thinning (a decrease of n in Eq. 13) is responsible for the growth of static zone of extrusion of plastic bodies. In order to study the dependence of the streamline on the consistency index, K, the shear thinning exponent was held constant at 0.5, as K is varied. Results

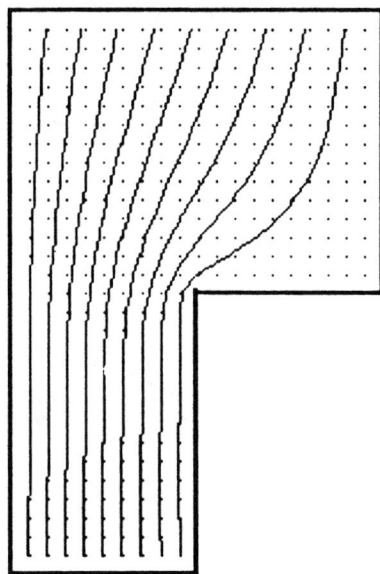

Figure 3 Streamlines of flow in a converging entry die. Viscosity obeys the power law with K=2 kPa·s^n, n=1.

indicate with this particular model, the streamlines for each case are nearly identical.

SUMMARY AND CONCLUSIONS

A finite difference method based upon the vorticity-stream functions was used to model the pseudoplastic flow behavior in a converging entry die and to calculate the flow velocities in electrical porcelain pastes. The flow patterns of behavior caused by the variation of the shear thinning exponent exhibit the full range of behavior between ideal plastic and Newtonian. This exponent is the most important parameter for these types of materials with low shear yield stresses. Decrease of the thinning exponent in materials having pseudoplastic behavior, leads to a larger static zone in the die entrance region.

REFERENCES

1. J.J. Benbow, T.A. Lawson, E.W. Oxley, and J. Bridgwater, "Prediction of Past Extrusion Pressure," *Am. Ceram. Soc Bull.*, **68** [10] 1821-1824 (1989).
2. J.J. Benbow, E.W. Oxley, and J. Bridgwater, "The Extrusion Mechanics of Pastes - The Influence of Paste Formulation on Extrusion Parameters," *Chem. Eng. Sci.*, **42** [9] 2151-62 (1987).
3. J.J. Benbow, "The Dependence of Output Rate on Die Shaping During Catalyst

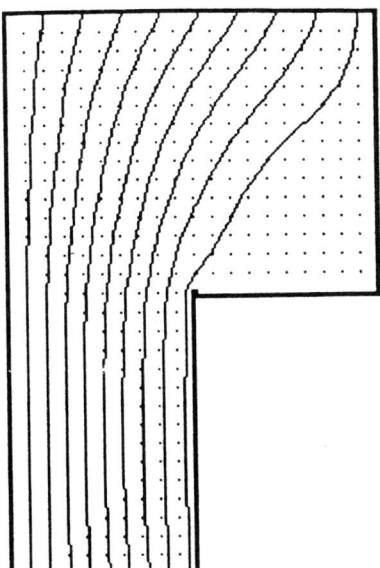

Figure 4 Streamlines of flow in a converging entry die. The viscosity obeys the power law with K=2 kPa·s^n, n=0.4.

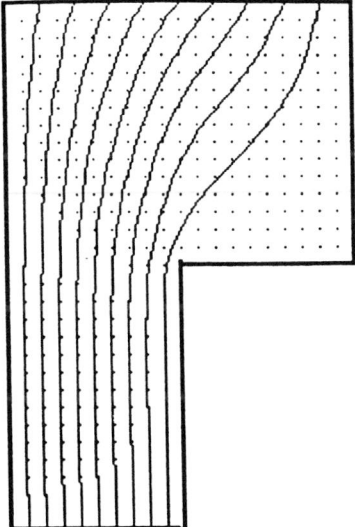

Figure 5 Streamlines of flow in a converging entry die. The viscosity obeys the power law with K=2 kPa·s^n, n=0.25.

Extrusion," *Chem. Eng. Sci.*, **26** 2151-62 (1971).

4. J.L. White, "Principles of Polymer Engineering Rheology," A Wiley-Interscience Publication, New York, 1990.

5. O.C. Zienkiewicz and P.N. Godbole, "Flow of Plastic and Visco-Plastic Solids with Special Reference to Extrusion and Forming Process," *International Journal For Numerical Methods in Engineering*, **8** 3-16 (1974).

6. O.C. Zienkiewicz, The Finite Element Method, 3rd Edition, McGraw-Hill Book Company (UK) Limited, London.

7. P.J. Roache, Computational Fluid Dynamics, Hermosa Publishers, Albuquerque, N.M. (1982).

8. J.S. Reed, Introduction of the Principles of Ceramic Processing, A Wiley-Interscience Publication, New York, 1988.

9. T. Martin, "Electrical Porcelain Extrusion: Analysis of Effect Due to Reprocessed Material Additions and Liquid Content Variations," Master's Thesis, Alfred University, 1990.

10. X.-L Luo and E. Mitsoulis, "A Numerical Study of the Effect of Elongational Viscosity on Vortex Growth in Contraction Flows of Polyethylene Melts," *J. Rheol.*, 34 [3] 309-43 (1990).

SUPERABSORBENT POLYMERS AS TEMPLATES IN FORMING OXIDE CERAMICS

Anne B. Hardy and Wendell E. Rhine
Ceramics Processing Research Laboratory, Massachusetts Institute of Technology, Cambridge, MA 02139

ABSTRACT

Superabsorbent polymers, a class of crosslinked polymers which absorb large quantities of water, were used to control the shape of a ceramic product in a modification of the relic process. Three oxide sources were used to impregnate the polymer: alkoxide hydrolysis, metal salts, and oxide sols. The infiltrated polymer was heated to remove the polymer, and the oxide retained the shape of the polymer. Because the polymers are highly absorbent, the amount of polymer necessary to maintain the shape was significantly reduced.

INTRODUCTION

In ceramics processing, control of morphology is often of critical importance. Because ceramics are solids with high melting points, many of the processing approaches used to control morphology in metals or polymers are not feasible. One method for forming ceramic fibers[1-5] or beads[6] is the relic process, where ceramics are formed by infiltrating a polymer with a precursor and then heating it to remove the polymer and convert the precursor to the ceramic form. This method incorporates the processing flexibility of polymers into the production of ceramics. In a standard formulation,[1] cellulosic fibers, such as rayon, imbibe an aqueous metal salt solution. The fibers are then heated to remove the polymer and convert the salt to the oxide. The oxide retains the shape of the original fiber. In addition to cellulose derivative fibers, a variety of novel precursor fibers have been used, including activated carbon fibers[2-4] and intercalated graphite fibers.[5]

Highly absorbent polymers, called superabsorbent polymers, have been developed which are capable of absorbing upwards of 100 mL H_2O/g polymer; the absorption depends on the polymer type and on the swelling conditions. The polymers are crosslinked and retain an expanded form of the original shape on swelling. After swelling, the polymer network consists primarily of water with the small amount of polymer maintaining the shape. Because many common ceramic processes are based on aqueous systems, superabsorbing polymers appear to be good candidates for use in conjunction with aqueous-based processing to control the shape of the ceramic product. Three different methods were used: alkoxide hydrolysis, metal salt decomposition, and gelation of oxide sols. Two superabsorbing polymers were used: Quat,[TM] a bead-shaped polymer with diameters ranging from 5 to 100 μm (dry), and Sanwet,[®] a superabsorbing polymer which was supplied in the shape of ~2.8 cm high animal figures.

EXPERIMENTAL PROCEDURE

Quat[TM*] was obtained commercially, and the following three methods were used to transform the polymeric beads into ceramic beads.

Alkoxide hydrolysis: The Quat polymer was swollen in water (pH 4) for 1 h. It was then filtered and rinsed with isoamyl alcohol to remove excess water from between the beads. The water-swollen beads were reacted for 20 h with tetraethyl orthosilicate (TEOS[†]) diluted 1:1 by volume with hexanes. The beads were filtered and rinsed with isopropyl alcohol.

Metal salt decomposition: An aqueous aluminum salt solution was prepared using $Al(NO)_3 \cdot 9H_2O$. The pH was raised to 2.5-3 using hexamethylenetetramine. The Al concentration was 3.6 wt%. Quat polymer was swelled in an excess of $Al(NO_3)_3$ solution for >2 h. The beads were isolated by filtering and then added to a stirred NH_4OH solution. The beads were filtered, then stirred in isopropanol, and then filtered again.

Sol gelation: Quat beads were swollen in three different ZrO_2 sols[‡]: ZrO_2 acetate (diameter = 5-10 nm), ZrO_2 50/20 (diameter = 50 nm) and ZrO_2 150/20 (diameter = 150 nm). The oxide concentration was 20 wt% for all three sols. The beads were left in the sol for >2 h and then separated from the sol by filtration. The ZrO_2 impregnated beads were added to isoamyl alcohol and stirred for 5-10 min. The beads were filtered and rinsed with isopropyl alcohol.

Sanwet,[®§] a starch grafted sodium polyacrylate, was supplied as animal shapes measuring approximately 2.8 cm high. The polymer was preswollen in deionized water (>3 days) and then placed in a ZrO_2 sol for 4 days. The ZrO_2-infiltrated polymer was then either placed in NH_4OH or exposed to NH_3 (g) to gel the sol. The polymer was dried under vacuum at 60-80°C for ~10 h before heating.

---

[*]   Bernard J. Obenski and Co., Berwyn, PA.
[†]   Aldrich Chemical Co., Inc., Milwaukee, WI.
[‡]   Nyacol Products, Ashland, MA.
[§]   Hoechst Celanese, Portsmouth, VA.

Both Quat and Sanwet polymers were removed by heating the samples in air at 2°C/min to 500°C followed by heating to 850°C at 5°C/min. Bead morphology was examined using a scanning electron microscope[*] (SEM). Bead density was measured with a helium pycnometer.[†] The densities of the pieces formed from Sanwet were measured by the Archimedes method.

RESULTS AND DISCUSSION

The focus of this project was to demonstrate that superabsorbing polymers could be used in conjunction with typical aqueous ceramic processing routes to control the shape of the ceramic product. Because superabsorbing polymers are much more absorbent than other preforms that have been used in the relic process, we also wanted to see if we could significantly reduce the amount of polymer needed to control the shape. Removal of large amounts of polymer is inefficient and can lower the quality of the ceramic product. Quat and Sanwet were used because they showed the effects of different polymer types with different absorbancies, and because they enabled us to change the size scale of the product by several orders of magnitude.

Figure 1 shows that for both Quat and Sanwet, absorbency clearly decreased with decreasing pH, although the polymers still absorbed significant amounts of water above pH 2. Quat was consistently more absorbent than Sanwet. Both Quat and Sanwet are ionic polymers and rely on charge repulsion between ions to aid in swelling. The presence of salts and changes in pH decrease the distance over which the ionic repulsion acts and can inhibit the polymer swelling. In addition, multivalent cations can sometimes crosslink between polymer hydroxyl groups, and this increase in crosslinking can limit the polymer absorbency.

Each of the processing methods used had constraints that determined the pH that was used. Hydrolysis of TEOS required either an acid or base catalyst. The more acidic the solution, the faster the reaction occurred. We chose a pH of 4 because the Quat was highly absorbent, and yet the reaction proceeded fast enough to give good silica yields within 20 h. Aluminum nitrate solutions were very acidic as prepared (pH < 1); however, the pH could be raised to about 4 before the solution began to gel. ZrO_2 sols rely on surface charge to prevent agglomeration of sol particles and the sols were most stable towards agglomeration at pHs of 3-3.5.

The Sanwet polymer was not compatible with TEOS or with many common organic solvents (e.g., hexane and toluene). On being added to TEOS it often crazed or split rapidly into smaller pieces. Also, due to difficulties drying the aluminum salt solutions in the large Sanwet samples, Sanwet was used primarily with ZrO_2 sols.

---

*   Model S-530, Hitachi, Ltd., Tokyo, Japan.
†   Quantachrome, Syosett, NY.

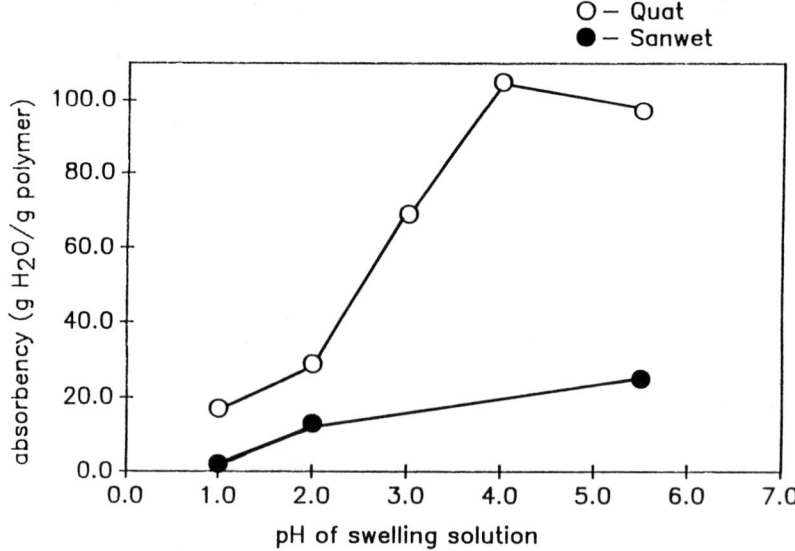

Figure 1 Absorbency as a function of pH for Quat and Sanwet.

Table 1 lists the ratio of the oxide product weight to the original polymer weight for the three methods. For comparison it also lists the product to polymer ratio for alumina fibers formed from rayon. The highest yields were obtained for silica beads, but even for the alumina beads the product yields per weight of polymer were greater than 400% that of the rayon. The low alumina yields compared to silica and zirconia were largely a reflection of the low aluminum concentration in the salt solution. By using superabsorbing polymers it was clearly possible to minimize the amount of polymer used.

ZrO_2 sols with different particle sizes were used to determine how the absorption was affected by particle size. For larger particle size sols (i.e., ZrO_2 50/20 and ZrO_2 150/20), essentially no ZrO_2 was incorporated into either the Quat or the Sanwet polymer. The larger sol particles were apparently too large to move through the open porosity in the swelled polymer. We also found with the large Sanwet samples that it was much more efficient to preswell the polymer in deionized water so that the polymer was fully expanded before absorbing the sol particles into the polymer.

The silica beads were translucent, and inspection of the interior of broken particles showed that even large particles were solid. Zirconia bead surfaces were cracked, and broken particles also had cracks in the particle interior. The alumina particles were less spherical in shape, probably in part because of the low oxide yield and

Table 1 Summary of Oxide Properties.

Polymer	Oxide Source	Metal Ion Conc. (wt%)	g Oxide/ g Polymer	Density (g/cc)	Oxide Bead Diam./ Polymer Diam.
Rayon[1]	$Al(NO_3)_3$	---	0.17-0.25	---	---
Quat	TEOS	---	14	2.1	1.9
Quat	$Al(NO_3)_3$	3.6	1.3	3.0	.76
Quat	ZrO_2 acetate	20	6	5.3	1.1
	ZrO_2 50/20	20	<0.1	---	---
	ZrO_2 150/20	20	<0.1	---	---
Sanwet	ZrO_2 acetate	20	0.8	2.7	---
	ZrO_2 50/20	20	---	---	---
	ZrO_2 150/20	20	---	---	---

hence the greater shrinkage on drying and heating. Representative SEM micrographs of the beads are shown in Figure 2. From these micrographs we estimated that the bead diameters after heating to 850°C ranged from approximately 10 to 300 μm. Because of the wide range in particle sizes it was difficult to estimate the overall change in diameter from polymer to oxide particle, but a ratio of the oxide particle diameter to the polymer particle diameter was calculated from the change in mass and density (we assumed the density of Quat was 1 g/cm³). Table 1 shows that the calculated ratios of oxide diameter to polymer diameter ranged from 1.9 for silica to 0.75 for the alumina beads.

The Sanwet polymer underwent large volume changes during the process, but overall the linear shrinkage from the original polymer to the oxide form was 5-10% (see Fig. 3). After removing the polymer, the pieces were not very strong, particularly in the interior, which was often very powdery and appeared less dense than the surface. The interior of the oxide pieces always had hollow regions which were due, in part, to hollow regions in the original polymer. However, the hollow regions in the product also appeared to result from the response of the water-swollen polymer to the sol dispersion. When the preswollen polymer was placed in the sol it began to shrink (the polymer absorbency in the sol was lower than in the water in which it was preswollen). As the polymer shrank it apparently reached a point at which the open space in the polymer became too small to admit the sol particles and infiltration was effectively stopped.

(a)

(b)

(c)

Figure 2 Scanning electron micrographs of oxide beads after heating to 850°C (a) SiO_2, (b) ZrO_2, and (c) Al_2O_3.

146

Figure 3 ZrO_2 ceramic formed from Sanwet polymer infiltrated with ZrO_2 acetate sol (after heating to 850°C).

CONCLUSIONS

A superabsorbing polymer was used in conjunction with alkoxide hydrolysis, salt decomposition, and gelation of oxide sols to form oxide beads. We showed that Quat, a highly absorbent polymer, could be used with a variety of techniques to significantly reduce the amount of polymer needed to control the shape compared to a traditional polymer such as rayon, and that the morphology was maintained well after heating. Sanwet was less versatile and ceramics formed from the less absorbent Sanwet tended to have hollow regions.

ACKNOWLEDGMENTS

The authors wish to thank AFOSR (Contract No. F49620-89-C-0102) for financial support.

REFERENCES

1. B.H. Hamling, U.S. patent 3,385,915, issued to Union Carbide, 1968.
2. R.J. Card, "Preparation of Hollow Ceramic Fibers," *Adv. Ceram. Mater.*, **3** [1] 29-31 (1988).
3. R.J. Card and M.P. O'Toole, "Solid Ceramic Fibers via Impregnation of Activated Carbon Fibers," *J. Am. Ceram. Soc.*, **73** [3] 665-68 (1990).
4. D.J. Waller, A. Safari, R.J. Card, and M.P. O'Toole, "Lead Zirconate Titanate

Fiber/Polymer Composites Prepared by a Replication Process," *J. Am. Ceram. Soc.*, **73** [11] 3503-506 (1990).

5. B.W. McQuillan and G. Reynolds, "Growth of Alumina Fibers from Intercalated Graphite Precursor Fibers"; pp. 739-45 in Ultrastructure Processing of Advanced Ceramics. Edited by J.D. Mackenzie and D.R. Ulrich. J. Wiley and Sons, New York, NY, 1988.

6. A.B. Hardy, W.E. Rhine, and H.K. Bowen, "Preparation of Porous Oxide Beads using Polymeric Beads to Control Bead Size and Shape"; pp. 1009-14 in Better Ceramics Through Chemsitry IV., Mat. Res. Soc. Symp. Proc. Vol 180. Edited by B.J.J. Zelinski, C.J. Brinker, D.E. Clark, and D.R. Ulrich. Materials Research Society, Pittsburgh, PA, 1990.

PROCESSING OF FOAMED CERAMICS

William P. Minnear
GE Corporate Research and Development, Optical and Electrical Ceramics
Program, P.O. Box 8, K-1, MB161, Schenectady, NY 12301

ABSTRACT

Porous ceramics for use in molten metal filtration, heat exchangers, etc. can be made directly by foaming slips with reactive prepolymers. Multiple regression analysis is used to establish the effects of temperature, pH, surfactants, foaming agents, H_2O content, and ceramic particle size on the foamed body characteristics: rise height, cell size and pore-interconnectivity. A model describing the interrelationship of these characteristics is presented.

INTRODUCTION

Low density ceramics and metals have a wide variety of applications ranging from molten metal filters, insulators, and heat exchangers to catalyst supports and abrasives.

Numerous process routes have been employed to fabricate porous structures. The most common is based on the replication of an existing organic preform by slurry infiltration and controlled expulsion—"dip and squeeze."[1-13] The dried slurry deposits sinterable powders on the preform. The organic is removed during a heat treatment step followed by sintering of the powders.

An alternative to dip-and-squeeze has been patented by Wood et al.[14] Their procedure employs reaction of an isocyanate capped polyoxyethylene polyol with the water of an aqueous powder slurry to directly form a porous body. The chemical reaction between the polymer and the water generates CO_2 which foams the viscous mass, and creates a reactive site for cross linking. The cross-linked hydrophilic foam serves as a carrier for the powder slurry. This technique offers several advantages including the variety of materials which can be incorporated into the slurry (fibers, etc.), the ease of manufacturing, and the extent to which cell size, cell wall integrity (reticulation) and total porosity can be varied.

This investigation is aimed at elucidating the effects of several process variables on the characteristics of the resulting foamed structures. Temperature, excess water, and surfactant were chosen as variables since they have been shown to affect the water-prepolymer reaction.[14] The alumina slip was acidified with HCl to improve fluidity. The effect of acid added beyond that required to deflocculate the slip was included as a variable. Acetone represents a class of additives which because of their high vapor pressure near room temperature could contribute to the "blowing" action of the CO_2. The heat of reaction would enhance any such effect. Acetone also serves to dilute the prepolymer, lowering its viscosity and possibly improving mixing. Finally, the powder particle size and surface area was varied to investigate possible contributions both to cell wall rupture from the presence of large agglomerates and to restriction of available water by surface adsorption.

EXPERIMENTAL

Two components were formulated separately to ultimately contain all of the ingredients for the trial. Component A included prepolymer and acetone. Component B contained: 1) distilled water, 2) alumina, 3) HCl, and 4) surfactant. All of the trials reported here utilized 8.1 g of prepolymer.* Two grades of alumina** were blended in various ratios to total 34 g: Al6SG, a fine submicron powder and Al4 a coarser agglomerated powder. Following are the ranges of independent variables: H_2O:11-17 g, Al4:0 to 34 g, surfactant†:0-0.5 g, T°(Comp B) 289-311K, pH:1-6(HCl†† 0.1-1.0 cc), acetone††:0-1 g.

The components were mixed separately by hand in 100 cc cylindrical polyethylene beakers approximately 4 cm in diameter under ambient conditions. Temperature was varied by chilling or heating Component B. Component B was added to the prepolymer mixture and stirred vigorously by hand using a metal spatula until a creamy consistency was achieved. Previous experience indicated that this specific mixing procedure did not affect the outcome of the trial. The mixture was allowed to rise, cure, and dry. Mixing time was typically 15 seconds; rise time 2-3 minutes; cure time (tack-free) 5 minutes; and drying time 24 hours in a laboratory hood. A total of 70 trials, spanning the variables and ranges reported, were used in this study. Numerous combinations outside the reported range were attempted and often led either to foam collapse or to rapid polymerization with little rise. These were excluded from the analysis.

The dried specimens were sliced lengthwise and characterized with respect to rise height, pore size, and degree of cell-wall openness. Figure 1 shows typical specimens. The rise height was estimated to the nearest millimeter averaging through

---

*   HYPOL 2000, W.R. Grace & Co., Lexington, MA.
**   Al6SG and Al4, Aluminum Company of America, Bauxite, AR.
†   Pheronic L-61 Polyol, BASF Wyandotte Corp., Wyandotte, MI.
††   Reagent Grade, J.T. Baker, Inc., Phillipsburg, NJ.

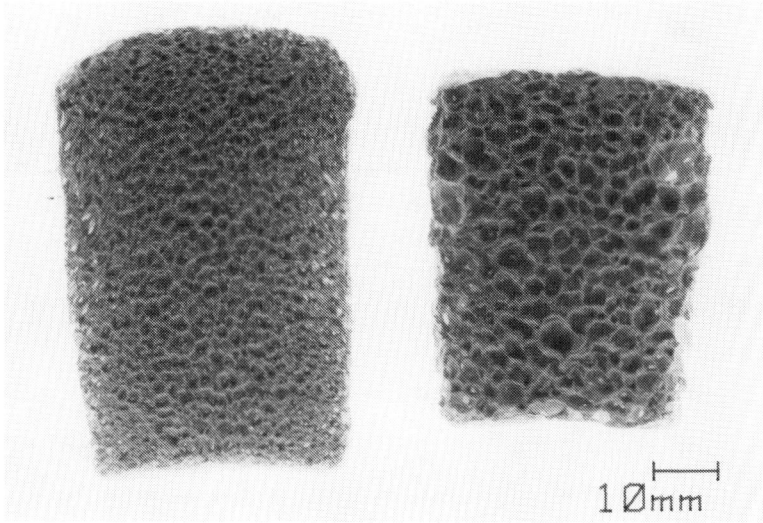

Figure 1 Examples of typical foamed specimen cross-sections.

the normally dome-shaped top. The pore size was determined on the cross section under low power (~5x) magnification by a linear intercept method. The degree of reticulation was established on a scale from 1 to 5 inclusive, also at ~5x magnification. A rating of "1" is defined as few small holes in cell walls, <10% of the wall area, with an occasional cell wall completely broken. A rating of "5" indicates very good reticulation with >90% of cell walls having many holes and many cell walls completely missing. A rating of "3" means that the walls between adjacent cells contained approximately the same amount of open and closed area.

RESULTS AND DISCUSSION

The foam height after drying ranged from 3.9 to 7.4 cm. The volume of the solids contained in the foam totals approximately 16 cm^3 and equates to a "foam height" of 1.2 cm. Therefore, even those trials with the least amount of rise contained appreciable porosity. For the cylindrical beaker used, the foam height is directly proportional to bulk density. The average pore size (R) varied from about 1 to 5 mm. Except for the pores adjacent to walls and free surfaces, the pore size was very uniform.

Multiple regression analysis was used to determine the effect of the independent process variables on the measured properties of the foam. Each of the three measured parameters, height, pore size and reticulation factor were regressed separately against a function of the water content, the H$^+$ concentration, the mass of surfactant, the absolute temperature, the mass of acetone, and the fraction of Al4

Table 1 Results of Regression Analysis of Process Variables Affecting Foaming.

	Water	pH	Surf	T °	Acetone	Powder Particle Size
Height	+	-	0	-	-	0
Reticulation	+	-	0	-	-	0
Pore Size	-	+	0	+	+	0

+	Positive Correlation
-	Negative Correlation
0	No Correlation

in the alumina mixture. A coefficient of variation (COV) was calculated for each independent variable by dividing the standard error of the coefficient by the coefficient. An arbitrary cutoff of 0.5 was chosen to determine significance. Therefore, if COV $\leq$ 0.5, then the process variable is reported to have an effect on the measured property. The results are summarized in Table 1.

The table shows that both the quantity of surfactant and the average powder particle size have no effect on foam rise, reticulation or pore size. Beyond that the table shows the difficulty in arbitrarily varying these three foam characteristics. For example, if the degree of reticulation of a particular formulation was to be increased, a change in only one process variable would necessarily result in a change of foam height and pore size.

Foam Constant Concept

Part of the interrelationship of foam height, pore size, and reticulation can be explained by the following argument based essentially on conservation of mass. Consider a cubic cell of side length R. A number of cells are contained in a rectangular volume of base L^2 and height H. The number of cells, $N = L^2H/R^3$. The volume of material, v, (polymer, slurry, etc.) contained in a single cell wall is approximately $6R^2(T/2)$ where T is the cell wall thickness. The factor of one-half results from wall sharing by two adjacent cells. The total material volume, V, contained in the volume is

$$N \times v = 3 \frac{L^2 HT}{R}. \qquad (1)$$

For constant volume and constant L, equivalent to allowing the foam to rise only in one dimension: HT/R = constant. In terms of the measured foam parameters: R = pore size, and H = rise height. If one assumes that the cell walls break at some characteristic thickness, then the reticulation factor, F, should be proportional to $1/T$. Since HT/R = constant, then

$$\frac{H}{FR} = \text{foam constant}. \qquad (2)$$

This relationship is consistent with the correlations presented in Table 1 and the illustrative example. Plotted in Figure 2 is the ratio of foam height to pore size (H/R) and the reticulation factor, F. This excellent linear relationship supports both the foam constant concept and the assumption that cell walls break at some characteristic thickness.

The Foaming Phenomenon

Cells grow from CO_2 bubbles which nucleate early in the mixing operation. Before significant cross linking occurs, cells would be free to interact and coarsen in response to forces driven by capillarity. As the average cell size grows, new nuclei would rapidly be incorporated into existing cells and would not contribute to the

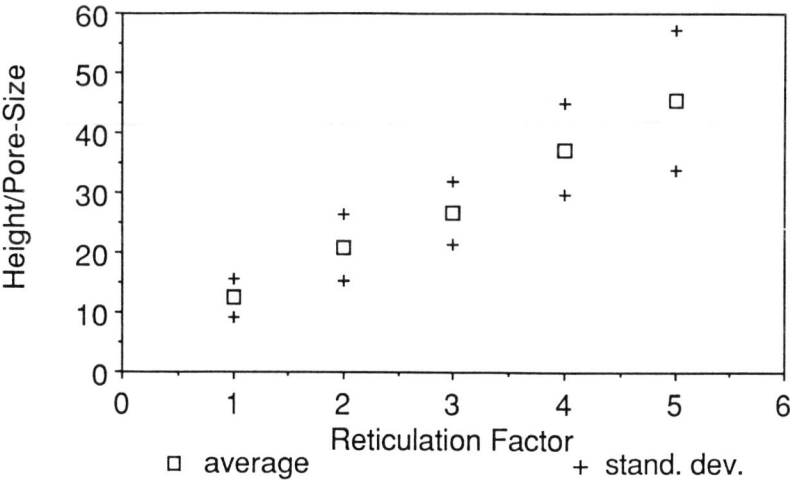

Figure 2 Height/pore-size vs. reticulation factor.

final pore size. As polymerization starts to occur, the semi-solid mass is able to support its own weight against the evolution of gas. During the rise, cell walls must remain substantially intact to prevent collapse. Therefore, the progress of the event could be described as follows: The number of pores is established early during mixing and is essentially constant during rise. Spontaneous reticulation occurs at or near the end of the rise by thinning of cell walls. The larger-pore-size foams actually emit audible "clicks" as the cell walls are breaking near or just after the end of rise. The wall thinning can result either from a small increase in cell size toward the end of rise or by drainage[15,16] at constant cell size.

Effect of Process Variables

The most obvious effect of adding acetone and increasing temperature is the reduction of the mixture viscosity which makes bubble coalescence easier during the nucleation stage. With fewer cells present at the start of rise, the average pore size in the foamed body would be larger and would correlate with temperature and acetone as shown in Table 1. On the other hand, increased water makes more reactant available to the prepolymer and should increase the number of nuclei, thus decreasing the pore size as shown.

The foam height depends only on the amount of gas evolved barring collapse. Cross linking cannot occur until water has first reacted with an isocyanate group to form CO_2. Excess water at the start of the reaction should therefore favor CO_2 evolution over cross linking and lead to increased foam height as shown in Table 1. Since it is likely that the water-isocyanate reaction and the cross-linking reaction have different activation energies, by inference from Table 1 higher temperature must favor polymerization.

The degree of reticulation represents the number of cell walls that are thin enough to rupture under the combined forces of gas pressure and surface tension near the end of rise. Trials that tend toward complete cell wall breakage would also tend to be eliminated from this data set since they would have collapsed during rise. Figure 3 shows a good inverse correlation between reticulation and pore size. Foams with finer pores tend to have more open structures. For a given foam height this is explained by the foregoing geometric arguments. Simply restated, smaller pores have more wall area with a fixed solids volume and therefore thinner walls on average. The data further shows both that the large agglomerates (>25 μm) intro-duced by the Al4 do not contribute to the mechanical rupture of cell walls and that the high surface area Al6SG does not interfere with cell-wall drainage.

SUMMARY AND CONCLUSIONS

The rise height, cell size, and cell interconnectivity of a foamed prepolymer were related to a number of independent process variables.

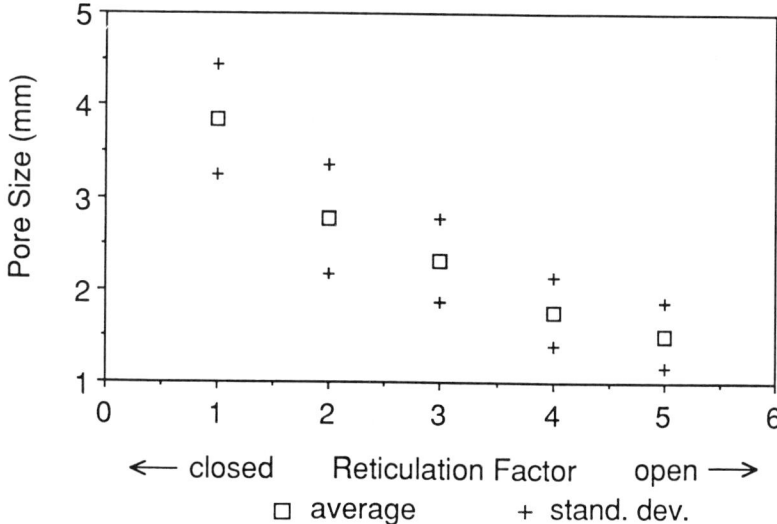

Figure 3 Pore size vs. reticulation.

The number of cells and the amount of CO_2 evolved are the two basic factors affecting all three foam characteristics. The latter determines the foam volume. Once the foam volume is established the number of cells determines the pore size and cell wall thickness. Cell wall rupture occurs when thinning reaches some critical value. The interdependence of these three factors is shown both by the data and a simple conservation of mass argument.

Powder particle size and surface area have no significant effect on foam characteristics. Acetone and temperature increase pore size by decreasing viscosity and encouraging cell coalescence early in the process. Excess water both provides more cell nuclei as well as increases foam height by supplying more reactant for the evolution of CO_2. The data suggest that pH has the same effect as temperature and acetone but no specific mechanism is suggested.

ACKNOWLEDGMENTS

The author wishes to acknowledge many useful discussions with Mark Miller regarding reaction chemistry and surfactants.

REFERENCES

1. K. Schwartzwalder and A. V. Somers, "Method of Making Porous Ceramic Articles," U.S. Pat. No. 3 090 094, May 21, 1963.
2. I. J. Holland, "Method of Making a Porous Shape of Sintered Refractory Materi-

al," U.S. Pat. No. 3 097 930, Jul. 16, 1963.

3. F. E. G. Ravault, "Production of Porous Ceramic Materials Through the Use of Foam Attacking Agents," U.S. Pat. No. 3 845 181, Oct. 29, 1974.

4. M. J. Pryor and T. J. Gray, "Molten Metal Filter," U.S. Pat. No. 3 893 917, Jul. 8, 1975.

5. F. E. G. Ravault, "Manufacture of Porous Ceramic Material," U.S. Pat. No. 3 907 579, Sept. 23, 1975.

6. J. C. Yarwood, J. E. Dore and R. K. Preuss, "Ceramic Foam Filter," U.S. Pat. No. 3 962 081, June 8, 1976.

7. C. Washbourne, "Catalyst Carriers," U.S. Pat. No. 3 972 834, Aug. 3, 1976.

8. M. J. Pryor and T. J. Gray, "Ceramic Foam Filter," U.S. Pat. No. 3 947 363, Mar. 30, 1976.

9. J. C. Blome, "Molten Metal Filter," U.S. Pat. No. 4 265 659, May 5, 1981.

10. J. W. Brockmeyer, "Ceramic Foam Filter," U.S. Pat. No. 4 343 704, Aug. 10, 1982.

11. J. W. Brockmeyer, "Ceramic Foam Filter and Aqueous Slurry for Making Same," U.S. Pat. No. 4 391 918, Jul. 5, 1983.

12. J. C. Yarwood, J. E. Dore and R. K. Preuss, "Method of Preparation of Ceramic Foam," U.S. Pat. No. 4 075 303, Feb. 21, 1978.

13. M. J. Pryor and T. J. Gray, "Method of Preparing Molten Metal Filter," U.S. Pat. No. 4 056 586, Nov. 1, 1977.

14. L. L. Wood, P. Messina and K. C. Frisch, "Method of Preparing Porous Ceramic Structures by Firing a Polyurethane Foam that is Impregnated with Inorganic Material," U.S. Pat. No. 3 833 386, Sept. 3, 1974.

15. A. W. Adamson, Physical Chemistry of Surfaces, 5th ed; pp. 546-550. Wiley, New York, 1990.

16. J. J. Bikerman, Foams, p. 137. Reinhold, New York, 1953.

QUICKSET™ INJECTION MOLDING OF HIGH PERFORMANCE CERAMICS

B.E. Novich, C.A. Sundback, and R.W. Adams
Ceramics Process Systems Corporation, 155 Fortune Boulevard,
Milford, MA 01757

ABSTRACT

The Quickset™ injection molding process is currently used to manufacture high performance complex-shaped ceramic components for heat engine and high density microelectronic applications. The Quickset process is described and the process capability is discussed for current applications.

INTRODUCTION

Recent advances in ceramics processing have enabled advanced ceramic materials to be used as prototypes in leading edge applications such as heat engine components and thin film compatible single chip and multi-chip microelectronics packages. However, full scale integration has been severely limited by the ability of the ceramic industry to produce advanced ceramic components at a cost and at a reliability consistent with the requirements of the application.

Low cost fabrication requires that complex-shaped ceramic components have net shape dimensions and that expensive processing steps are not required to achieve specified properties; extensive component machining and process steps such as HIPping should be avoided. Achievement of net shape dimensions and ultimate material properties require that the green microstructure development be controlled at the particulate level. Process monitoring ensures that control is maintained at each processing step.

The Quickset™ process is a low pressure, low viscosity closed cavity forming process with demonstrated net shape capability for production of complex-shaped ceramic and metal components. The Quickset™ tooling and process costs are low relative to competitive forming technologies, so that component manufacturing is cost effective at both low and high volume manufacturing levels. The Quickset™

process flow is described below and each process step is highlighted. Finally, Quickset™ process capability is described for current structural and microelectronic applications.

EXPERIMENTAL PROCEDURE

The Quickset™ Process

The Quickset™ process flow chart is given in Figure 1. The primary process objectives are to: 1) form a homogeneous suspension by dispersing the powder uniformly in the pore fluid, 2) maintain the suspension homogeneity through each subsequent unit operation, and 3) process this suspension in a closed and clean environment to avoid introducing property-limiting defects. Components processed based on these objectives have sintered material properties approaching theoretical material properties and superior dimensional tolerances.

Mixing and filtration

In the mixing operation, pore fluid is combined with ceramic powder in a high shear

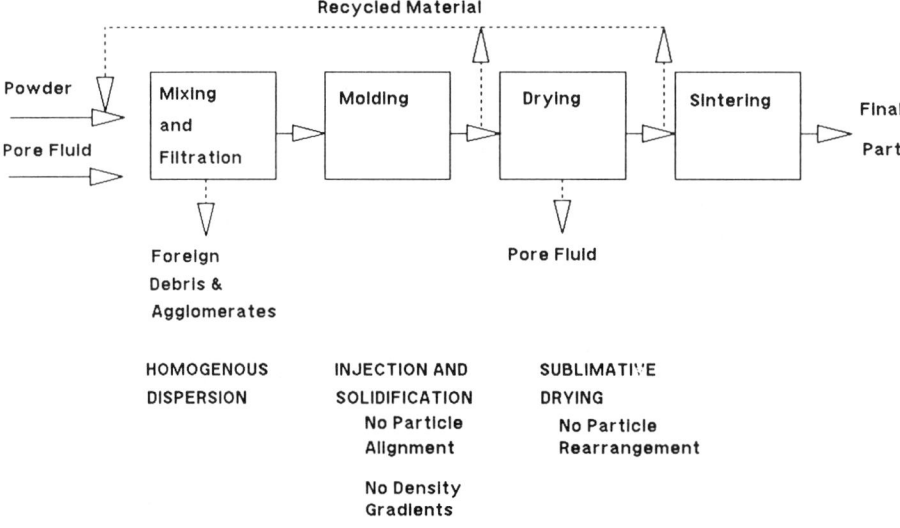

Figure 1 Quickset™ Process Flow Chart.

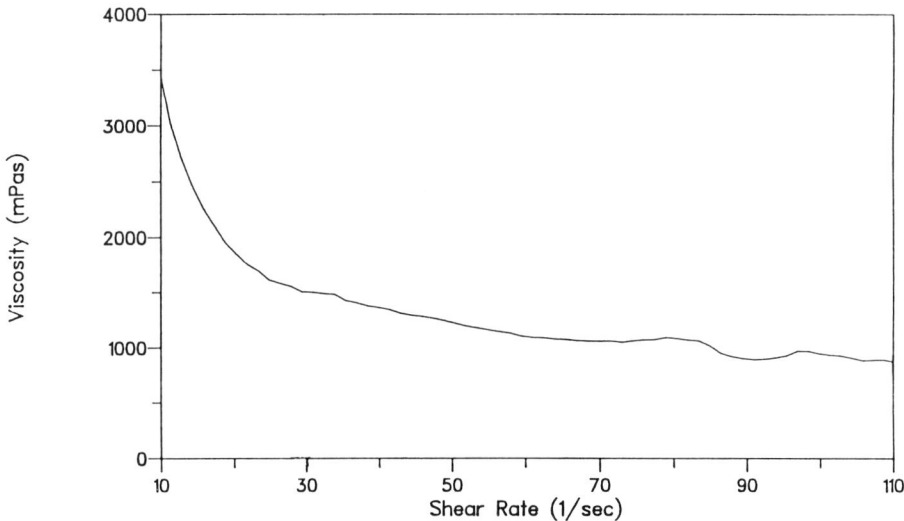

Figure 2 Rheology of Syalon 101 Quickset™ Feedstock Suspension. (Syalon 101 supplied by Vesuvius Zyalons.)

mixer in a filtered air environment. The pore fluid consists of an aqueous or non-aqueous liquid vehicle and dispersants. Prior work at Ceramics Process Systems[1,2] has demonstrated that single phase and composite ceramic suspensions can be produced with near-theoretical particle loadings and with fluid rheologies. Volume fraction solids in these suspensions can be greater than 55%. The suspension rheologies are Newtonian and the viscosities are typically lower than 2000 mPa at 100 sec[-1] shear rates. A typical viscosity curve is shown in Figure 2 for a submicrometer SiAlON suspension; the SiAlON powder has a specific surface area of 14 m^2/g and a mean particle size of 0.5 μm. After mixing, the suspension is introduced into a closed environment, degassed, and then pumped through a filter to remove any residual agglomerates or impurities. The suspension contacts only abrasion-resistant polymer during processing, eliminating metal contamination.

Injection molding

During the molding operation, the filtered suspension is injected into a closed cavity mold at a controlled rate. Injection pressures are typically less than 0.7 MPa; consequently tool clamping forces are low and soft tooling can be employed. The solidification of the pore fluid is controlled and rapid; a cryptocrystalline non-ceramic pore fluid phase is produced when the optimal additives and solidification rates are used. The volume change of the suspension during solidification is minimal and is uniformly controlled throughout the part. Minimizing the volume change during solidification yields a solidified particulate matrix which is as uniform as the original feedstock suspension. The solidified part is rapidly ejected from the mold and stored until the sublimative drying step.

Sublimative drying

Sublimative drying of ceramic powders is widely practiced.[3] However, sublimative drying of ceramic components has been applied infrequently;[4,5] sublimative drying of components has been practiced in applications where tight dimensional tolerances were not required and microstructural defects were acceptable.

The pore phase is removed from the molded part by sublimation. The solid pore fluid is converted directly to a gas without the formation of a liquid phase. During sublimation, capillary stresses associated with removing liquids from finely porous solids are avoided; these stresses develop during binder pyrolysis and evaporative drying operations where liquid phases are formed. Without a liquid phase, the particle matrix remains homogenous during drying because particles do not shift nor creep. Because particle rearrangement does not occur, nondestructive sublimation can be done more rapidly than nondestructive binder pyrolysis or evaporative drying.

Intelligent processing

Adams et al.[6] discussed the application of intelligent process control to Quickset™ technology. A description of the Quickset™ intelligent controller was given. The intelligent controller performs process monitoring and control, including real-time modification of process parameters for each unit operation. The key control variable for component dimensional tolerance is slip solids content which can be controlled to less than ±0.03 volume%. Component strength is strongly dependent on degassing time because voids from residual air bubbles are the dominant defect feature in Quickset™ pressureless sintered microstructures.

DISCUSSION

Process Capability

Described below are current applications in which Quickset™ technology has been applied to produce components with net shape dimensions and high performance material properties.

Ceramic rings

Syalon 101 rings were fabricated with sintered average dimensions of: 19.60 mm inner diameter, 40.91 mm outer diameter, and 25.00 mm height. Four-point flexural strength properties of Syalon 101 rings produced by Quickset™ injection molding are significantly improved over flexural strength properties of Syalon 101 rings produced by isopressing; Syalon 101 powder was supplied by Vesuvius Zyalons.

Forming Method	Flexural Strength	Weibull Modulus
Isopress	785 MPa	8.7
Quickset[TM]	975 MPa	19.4

The as-sintered averages and 3-sigma dimensional tolerances of one production lot of 30 Quickset[TM]-processed rings were:

Inner Diameter:	19.60 ± 0.08 mm
Outer Diameter:	40.91 ± 0.09 mm.

Heat engine components

Quickset[TM] fabrication of 14-pocket gasifier vane seat platforms and 26-blade axial flow gasifier turbine rotors is currently being performed under contract to the Allison Division of General Motors Corporation (Advanced Turbine Technology Applications Program sponsored by the U.S. Department of Energy). GM-Allison[7] reported significantly higher flexural strengths for Quickset[TM]-processed silicon nitride than for pressure slip cast-processed silicon nitride.

Shown in Figure 3 is an engine-ready gasifier vane seat platform which was fabricated using Quickset[TM] technology. The 4-point flexural strengths of testbars taken from platform hubs were:

Temperature (°C):	25	1000	1200	1300
Strength (MPa):	703	552	503	462

Duophase SiAlON[8] gasifier vane seat platforms were Quickset[TM]-processed and pressureless sintered in 20 lots of 10 components. Dimensional tolerance analysis was carried out on the 200 as-sintered vane seat platforms. The 3-sigma dimensional tolerances are listed below:

Platform Diameter	± 0.108 mm Within Lot
	± 0.110 mm Lot to Lot
Vane Pocket Cord Length	± 0.026 mm Within Part
	± 0.091 mm Within Lot
	± 0.113 mm Lot to Lot

Cavity chip carriers

Aluminum nitride multicavity substrates for high density interconnection are currently being manufactured using the Quickset[TM] process.[9] The key product requirements are high thermal conductivity, smooth surface finish and tight dimensional tolerances of the cavity features. These requirements are easier to achieve with Quickset[TM] processing because sintered microstructures have fewer pores and defects and are more homogeneous than sintered microstructures obtained with conventional processing. The thermal conductivity value is 230 ± 21 W/m-K (3-sigma) for Quickset[TM]-processed and pressureless sintered material; thermal conductivity values of 170 to 180 W/m-K are reported in current literature for

Figure 3 Quickset™ Injection Molded AGT-5 Gasifier Vane Seat Platforms. (Parts fabricated under contract to Allison Gas Turbine Division of General Motors Corporation for the ATTAP program.)

conventionally processed aluminum nitride.[10] Occhionero and White[9] reported aluminum nitride surface finishes of: <0.41 µm for as-molded, as-sintered surfaces and <0.05 µm for polished surfaces. Defect sizes for polished surfaces were <25 µm at typical defect densities of <1 per square centimeter.

Occhionero and White reported as-sintered cavity tolerances for 9 cavity substrates formed using the Quickset™ process; the cavity substrate is shown in Figure 4. Test part cavity size was approximately 11.43 mm X 11.43 mm with a cavity depth of 0.64 mm. As-sintered dimensional tolerance was determined for 21 test parts sampled from 3 production lots. Three-sigma dimensional tolerances were ±0.058 mm on cavity size and ±0.061 mm on cavity center-to-center position.

CONCLUSIONS

In comparison with conventional processing, complex-shaped components fabricated with the Quickset™ process have the following attributes:

- improved mechanical properties because of the use of clean closed-system processing and the uniformity and the defect-free nature of the sintered microstructure;

162

Figure 4 Quickset™ Injection Molded Aluminum Nitride Cavity Microelectronic Chip Carriers.

- improved component dimensional tolerances because of the maintenance of the homogeneously dispersed particle matrix during processing and the integration of intelligent control into the manufacturing process;

- lower component costs because of the minimization of secondary machining operations and the avoidance of expensive processing operations.

REFERENCES

1. B.E. Novich and M.A. Occhionero, "Pourable Suspensions of Particulate Materials", U.S. Patent #4,904,411.
2. B.E. Novich and J.W. Halloran, "Liquefaction of Highly Loaded Composite Particulate Ceramic Suspensions", U.S. Patent #4,882,304.
3. D.W. Johnson, "Innovations in Ceramic Powder Preparation", in Advances in Ceramics: Ceramic Powder Science, vol. 21, Edited by G. Messing, American Ceramic Society, 3-19 (1987).
4. G. Weaver and B. Nelson, "Molding Refractory and Metal Shapes by Slip-Casting", U.S. Patent #4,341,725.
5. N. Takahashi, "Method for the Freeze-Pressure Molding of Metallic Powders",

U.S. Patent #4,740,352.

6. R.W. Adams, W.B. Householder and C.A. Sundback, "Applicability of Quick-set™ Injection Molding to Intelligent Processing of Ceramics", to be published in the Proceedings for the Annual Conference on Composites, Materials and Structures, American Ceramic Society, Cocoa Beach, FL, January 16-18 (1991).

7. H.E. Helms, P.J. Haley, L.E. Groseclose, S.J. Hilpisch and A.H. Bell, "Advanced Turbine Technology Applications Project", to be published in the Proceedings of the 28th Automotive Technology Development Contractors' Coordination Meeting, Dearborn, MI, October 23-26, Society of Automotive Engineers (1991).

8. B.E. Novich, R.R. Lee, G.V. Franks and D. Ouellette, "Fabrication of Low Cost and High Performance Ceramic Gas Turbine Engine Components", to be published in the Proceedings of the 28th Automotive Technology Development Contractors' Coordination Meeting, Dearborn, MI, October 23-26, Society of Automotive Engineers (1991).

9. M.A. Occhionero and P.A. White, "Aluminum Nitride Cavity Substrates by Quickset™ Injection Molding", Proceedings of the 1990 International Electronics Packaging Conference, Marlborough, MA, September 10-12, IEPS, 1017-1025 (1990).

10. A. Mohammed, A. Abdo, G. Scarlett and F. Sherrima, "Effect of Lot Variations on the Manufacturability of Thick and Thin Film AlN Substrates", Proceedings of the 1990 International Symposium on Microelectronics, Chicago, IL, October 15-17, ISHM, 7-12 (1990).

ALCOHOL BASED BINDER SYSTEMS
FOR MOULDING CERAMIC MATERIALS

Mohan J. Edirisinghe, Katherine L. Tomkins and Michael Patching
Department of Materials Technology, Brunel University, Uxbridge,
Middlesex, UB8 3PH, U.K.

ABSTRACT

Alcohol-water and alcohol-ester binder systems have been used to make mouldable concentrated ceramic formulations. The rheology of the formulations were investigated and 25mm diameter cylinders were compression moulded and subjected to pyrolytic binder removal prior to sintering. Evaporation of binder during processing operations is a major problem and causes defects in the sintered samples.

INTRODUCTION

Injection moulding submicrometer size ceramic powders using the flow properties of different binders is an important technique for near-net shape forming engineering components.[1,2] Waxes and high molecular weight polymers have been favoured as binders[3] but their removal requires carefully controlled, slow pyrolysis (2-5°C hour[-1]) up to about 500°C[4] and is a major cause for high scrap rates especially in thick section (>10-15mm) components.[1,5] Therefore, the development of binder systems which could be removed by rapid pyrolysis at a low temperature is desirable.

The use of an alcohol as a binder in ceramic shape forming is not unknown. 2-methylpropan-2-ol[6] and octadecanol[7] have been used as the major binder in ceramic moulding processes. Water has also been used as the major binder.[8,9] However, use of 2-methylpropan-2-ol instead of water offers many advantages. It freezes at about 25°C and the use of cold water is sufficient to control solidification in the mould. The number of moles per unit volume of the alcohol is <20% when compared with that of water and therefore a significantly smaller amount of alcohol vapour is produced during debinding, thereby reducing the binder removal time. Also, in the case of the alcohol, alkyl, rather than OH[-] groups, adsorb on to the ceramic particles making desorption during binder removal easier.[6] However, even at 25°C the vapour pressure of 2-methylpropan-2-ol is approximately double that of water causing high evaporation losses which makes processing difficult.

Therefore, an attempt has been made in this research to use 2-methylpropan-2-ol -water and 2-methylpropan-2-ol-ester binder systems to process A16.SG alumina suspensions. Esters have been used as minor binders in several injection moulding formulations.[3]

EXPERIMENTAL DETAILS

Materials

Standard A16.SG alumina (ex Alcoa) with a very narrow particle size distribution (median 0.4 μm) and a specific surface area of 9.5 m^2g^{-1} was used as the ceramic powder. 2-methylpropan-2-ol (density 784 kgm^{-3}, ex BDH Chemicals, Poole, U.K.) and 100% pure sunflower oil (density 939 kgm^{-3}, ex Princess Foods, Liverpool, U.K.) were used as the alcohol and ester respectively. A 2:1 wt. ratio of alcohol:water (or ester) was used in the binder systems.

2-methylpropan-2-ol containers were heated in warm water to obtain it in the liquid state. Subsequently, the required amount of water or ester were added to prepare the binder systems which were immediately used for compounding as described below. The thermogravimetric weight loss data of the individual components and the two binder systems in static air was determined using a Perkin-Elmer TGS-3 thermobalance.

Compounding

The powder (91% by wt.) and the binder systems prepared were manually mixed in a glass container and then compounded at room temperature using a twin roll mill to produce 2kg batches. The formulations were cooled to 5°C immediately after compounding and kept at this temperature until used in the other experiments described below. Samples were ashed at 600°C to calculate evaporation losses in the formulations during compounding. Also, the weight loss in the compounded formulations at an ambient temperature of 20°C due to further evaporation was monitored.

Rheology

A Davenport capillary rheometer with a 1.5mm diameter, 35mm length die was used to measure the apparent viscosities of the formulations at several shear rates. The temperatures used were 30 and 50°C for the alcohol-water and alcohol-ester formulations respectively. The rheometer gives ±1°C temperature control even at these low temperatures.

Binder Removal and Sintering

25mm diameter and 30mm thick cylinders were compression moulded using a model M1/R hydraulic press (ex Apex Construction Ltd., Dartford, U.K.). A pressure of 12MPa at room temperature (approx. 20°C) was used. The cylinders were X-ray radiographed to verify that there were no voids or cracks present. They were then subjected to binder removal by placing in an oven at 90°C for 2 hours. Subsequently, they were sintered by heating to 1600°C. The cylinders were X-ray radiographed prior to and after sintering.

RESULTS AND DISCUSSION

Thermogravimetry

Results obtained are shown in Fig. 1. Evaporation in the alcohol-water binder system occurred over a slightly higher temperature range compared with that of the pure alcohol. However, compared to pure water, evaporation is still rapid. In fact, thermogravimetric studies on a 1:1 (wt. ratio) alcohol:water binder system (Fig. 1) show that by increasing the amount of water present evaporation could be made to occur over a larger temperature range which is more acceptable for ceramic moulding. The alcohol-ester binder system clearly showed an initial weight loss

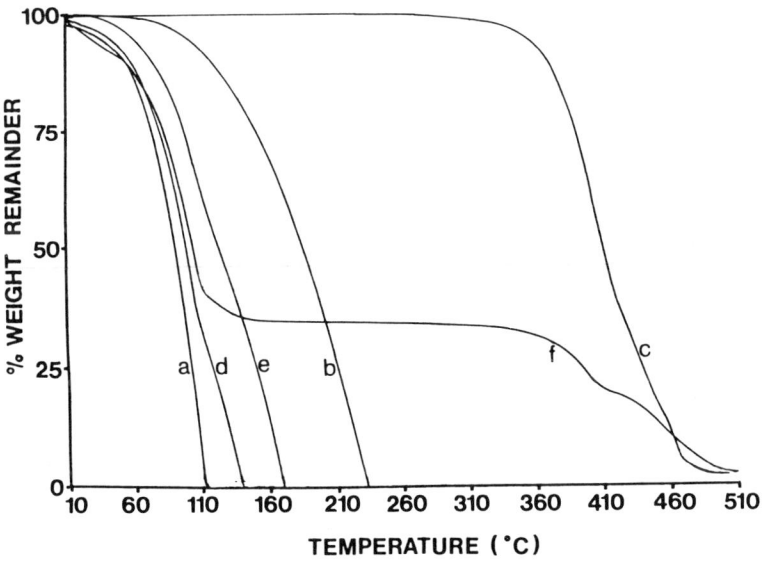

Figure 1 Thermograms of (a) the alcohol, (b) water, (c) the ester, (d) 2:1 alcohol-water binder system, (e) 1:1 alcohol-water binder system and (f) 2:1 alcohol-ester binder system, at a heating rate of 20°C hour^{-1} in static air.

167

due to the evaporation of the alcohol. This was followed by a period of inactivity until the ester was removed at temperatures in excess of 350°C but by then adequate porosity will be present in the moulded component for diffusion of the evaporating species and therefore this could be a part of the sintering operation.

Formulations

91% wt. ceramic in both formulations is equivalent to 70%vol. solids loading in the ceramic-binder binary system. However, the maximum ceramic volume fraction obtained from semi-empirical equations, which relate the viscosity of A16.SG alumina suspensions to %vol. of powder, is 0.73-0.76.[10] The critical powder volume concentration (CPVC), which is an upper limit of the ceramic volume loading during dynamic mixing, is 0.65 for this powder.[11] Therefore, it is possible that in these formulations all the voids between ceramic particles are not filled with binder as in a typical injection moulding system.[3] Ashing experiments on compounded samples revealed that the %wt. ceramic had increased to 93.2% and 95.8% respectively in the alcohol-water and alcohol-ester compositions respectively and therefore the actual formulations used in moulding operations were indeed a ceramic-binder-air system.

The alcohol-ester binder system showed phase separation when allowed to stand at ambient temperature. Also, thermogravimetric studies (Fig. 1) show a significant difference in volatility between the alcohol and ester. These two factors contributed to the faster weight loss observed in the alcohol-ester formulation, in comparison with the alcohol-water composition, when held at ambient temperature.

Rheology

Exploratory experiments revealed that, in terms of acceptable fluidity 30 and 50°C, for the alcohol-water and alcohol-ester formulations respectively, were the most appropriate temperatures to assess the rheology. Viscosity-shear rate graphs (Fig. 2) show a negative gradient indicating pseudoplastic behaviour which is essential for ceramic moulding operations.[3] However, in both formulations, there is a steeper increase in viscosity at lower shear rates. An important performance criterion of mouldable ceramic-binder formulations is the viscosity at a shear rate of about 100 s^{-1}.[12] Extrapolation of graphs in Fig. 3 shows that this value is approximately 200 and 100 Pa.s for the alcohol-water and alcohol-ester formulations respectively and compares favourably with 257 Pa.s for an injection mouldable 55%vol. A16.SG alumina-polypropylene formulation[10] at its processing temperature.

Binder Removal

X-ray radiographs of as moulded and binder removed components did not show any macro defects. However, cracks, most of them perpendicular to the axis of the cylinder, were present on the surface after sintering (Fig. 3). This defect was much

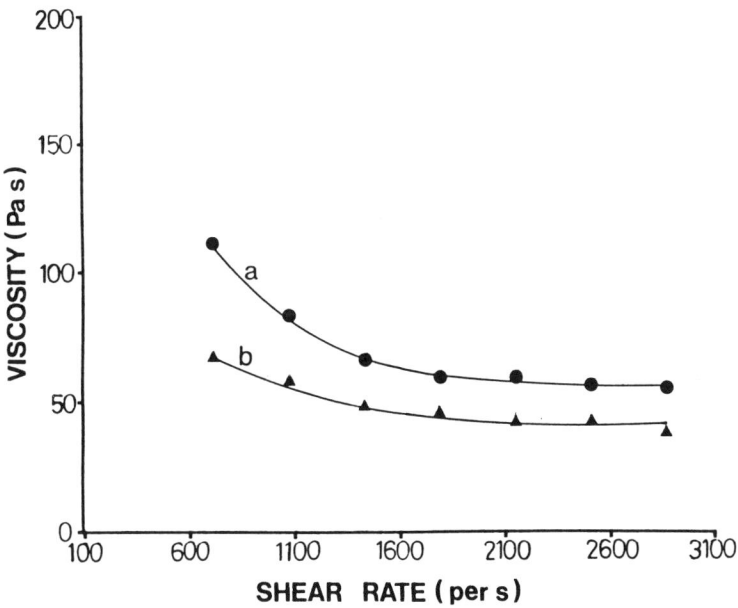

Figure 2 Viscosity vs. shear rate graphs of (a) 2:1 alcohol-water and (b) 2:1 alcohol-ester formulations at 30 and 50°C respectively.

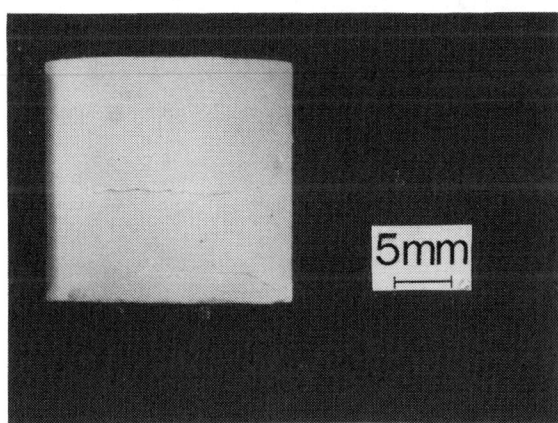

Figure 3 Appearance of sintered cylinder made using the 2:1 alcohol-water formulation.

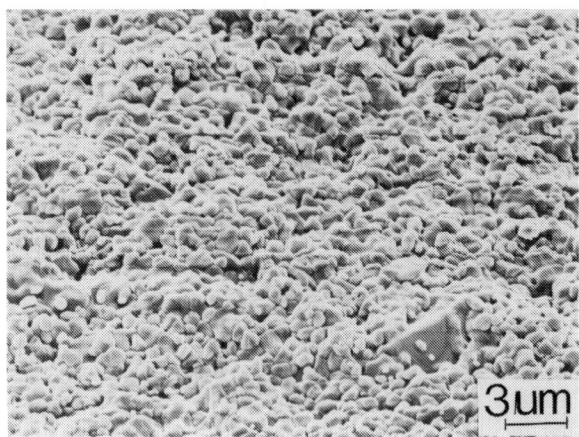

Figure 4 Scanning electron micrograph of the cross section region adjacent to the crack in Figure 3.

more evident in the cylinders made from the alcohol-water formulation. SEM studies (Fig. 4) revealed that the cross-section regions adjacent to these cracks were poorly sintered compared to other regions in the interior of the specimens. These defects are caused by the preferential evaporation of binder from the surface regions of the cylinders during moulding which reduces flow and packing of ceramic particles in these areas. The lubricating effect of the ester reduced this effect, although alcohol evaporates faster in the alcohol-ester formulation, and therefore fewer surface cracks were present in the cylinders made from this formulation.

CONCLUSIONS

The alcohol based binder systems investigated have acceptable flow properties for ceramic moulding operations. A major proportion of the binder can be successfully removed from the moulded components by rapid heating at temperatures <100°C. However, evaporation of binder during processing and especially moulding inhibits efficient and uniform packing of ceramic particles, even though these operations were carried out at the ambient temperature, and this leads to defects caused by differential sintering.

ACKNOWLEDGMENTS

The authors wish to thank Mr Harry Andrews for technical help and Mrs Kathy Goddard for typing the manuscript.

REFERENCES

1. R. Carlsson, "Shaping of Engineering Ceramics", Materials and Design, 10, 10-14 (1989).
2. M.J. Edirisinghe, "Injection Moulding of Ceramics", Metals and Materials, 6, 367-370 (1990).
3. M.J. Edirisinghe and J.R.G. Evans, "Review: Fabrication of Engineering Ceramics by Injection Moulding I. Materials Selection", Int. J. High Tech. Ceram., 2, 1-31 (1986).
4. J.K. Wright, J.R.G. Evans and M.J. Edirisinghe, "Degradation of Polyolefin Blends Used for Ceramic Injection Moulding", J. Amer. Ceram. Soc., 72, 1822-1828 (1989).
5. J.R.G. Evans and M.J. Edirisinghe, "Interfacial Factors Affecting the Incidence of Defects in Ceramic Mouldings", J. Mater. Sci., 26, 2081-2088 (1991).
6. T. Miyashita, H. Nishio, Y. Ueno and S. Kubodera, "Method of Moulding Powder Materials", U.K. Patent 2163 780, May 1987.
7. T. Sasaki, S. Yasuhara and H. Nishikawa, "Binder Compositions for Ceramic Mouldings", Japanese Patent 62278160, Dec. 1987.
8. M Kuwabara and M. Inoue, "Molding Material for Ceramic Articles", Japanese Patent 63265849, Nov. 1988.
9. T. Nakagawa, L. Zhang, H. Noguchi, N. Takahashi and K. Suzuki, "Compression Moulding of Fine Ceramic Powder by Using Water Binder", Mod. Dev. Powder Metall., 20, 763-772 (1988).
10. J.K. Wright, M.J. Edirisinghe, J.G. Zhang and J.R.G. Evans, "Particle Packing in Ceramic Injection Moulding", J. Amer. Ceram. Soc., 73, 2653-2658 (1990).
11. C.J. Markhoff, B.C. Mutsuddy and J.W. Lennon, "Method for Determining Critical Powder Volume Concentration in the Plastic Forming of Ceramic Mixes", in Advances in Ceramics, 9, 246-250 (1984), J. Mangels and G.L. Messing (Eds.), Amer. Ceram. Soc.
12. M.J. Edirisinghe and J.R.G. Evans, "The Rheology of Ceramic Injection Moulding Blends", Br. Ceram. Trans. J., 86, 18-22 (1987).

PROCESSING OF A HIGH TOUGHNESS SILICON NITRIDE MATERIAL

B. J. Meenan, R. A. Haber, and D. E. Niesz
Rutgers University, Piscataway, NJ 08855-0909

ABSTRACT

Parameters were examined for the aqueous processing of a silicon nitride powder doped with yttria, magnesia and silica. Materials were prepared by 1) co-mixing and 2) prereaction of individual oxide and nitride powders. Prereaction was found to yield composite particles whose mean particle size increased from 0.2 μm to 1.1 μm. A low viscosity (175 mPas) aqueous slip with solids loading of 38 volume percent was achieved with prereacted material, compared with the co-mixed powders whose viscosity at 34 volume percent loading exceeded 4000 mPas.

INTRODUCTION

The attractiveness of silicon nitride, over that of metals or traditional ceramics, comes from superior mechanical and thermal properties, coupled with its low specific gravity. Due to the inherent brittle nature of ceramic materials, much research has been conducted to increase the toughness of silicon nitride, without compromising its high temperature strength or sinterability. These efforts have produced high toughness material through the use of several additives in small percentages.[1]

Exceptional mechanical properties have typically been produced through hot pressing silicon nitride with minor amounts of additives or by incorporating second phase additions such as silicon carbide.[2] Recently a "self-reinforced" silicon nitride was found to provide high toughness and high strength without requiring second phase additions. Pyzik et al.[3] showed that by adding small percentages of up to four oxides to a fine grained silicon nitride, needle-like grains form, providing reinforcement to the silicon nitride matrix.

Producing homogenous aqueous dispersions of several oxide and nitride components is difficult due to mismatches in the surface charges of the different particle chemistries. In addition, it has been shown that silicon nitride powder reacts with water, forming ammonia gas and a silica coating on the particles. De Jong reported a shifting in the isoelectric point of a sub-micron silicon nitride powder until this reaction was complete.[4]

The purpose of this study is to produce a multi-component silicon nitride-based powder, similar to that studied by Pyzik, suitable for aqueous processing. In doing so, two powder processing procedures were selected. First, conventional mixing was chosen where three oxides were blended with silicon nitride, dried, slip cast into samples, and fired. In the second process the blended powders were subjected to an initial prereaction. In this process the powders are fired to temperatures where the oxide additives melt and coat the silicon nitride (a more complete overview of the prereaction process is provided by Cohn[5]). This material is then cooled, where the composite particles are ground to a desired particle size distribution, slip cast into samples, and fired. A comparison will be shown of the dispersion characteristics of the two processes.

EXPERIMENTAL PROCEDURE

Materials used throughout this study were SN E-10 silicon nitride,[*] Y_2O_3,[†] MgO,[‡] microcrystalline SiO_2,[§] reagent grade methanol,[¶] and Darvan C.[**] All water used was distilled and deionized.

To prepare powders for the prereaction process, silicon nitride powder and the desired additives were suspended in methanol (at 20 volume percent solids) and ball-milled with zirconia media. Once the media was removed, the mixture was dried at 43°C for at least 24 hours.

Dried powders were added to water and mixed in a high shear blender and subsequently ball-milled for up to 12 hours. Darvan C was added as needed to maintain low viscosity. To produce material for prereaction, the dried powder was pressed at 138 MPa into thin sheets, which were then reduced in size to -10/+20 mesh.

All reactions were carried out in an ASTRO furnace[††] equipped with a tungsten mesh heating element. A 5 cm diameter graphite crucible was filled with the pressed granules (approximately 65 rams) and capped with a graphite lid. A positive nitrogen flow of 20 cc/min. was used throughout each furnace cycle. The firing cycle

---

[*]   UBE Industries, Ltd.
[†]   CERAC, Inc.
[‡]   Aldrich Chemical Co.
[§]   Illinois Minerals Co.
[¶]   Fisher Scientific.
[**] R.T. Vanderbilt Co.
[††] Thermal Technology, Inc.

consisted of a 25°C/min. ramp to 1425°C where the temperature was held for 5 minutes and cooled at a similar rate.

Reduction of prereacted granule size was accomplished using a jet mill.[*] An air pressure of 240 MPa was used for each of the opposing jet streams.

Particle size and morphology was determined using three separate techniques. 3 volume percent of the jet-milled powder was suspended in water and dispersed with a Darvan C solution. An X-ray Sedigraph was used to determine the particle size distribution of milled material. This same solution was dried and examined using a scanning electron microscope. Dispersion characteristics were determined using a concentric cylinder viscometer where the shear rate varied from 1-200 s^{-1}.[†]

RESULTS AND DISCUSSION

For each processing procedure a nominal batch composition of 95:2:2:1 weight percent $Si_3N_4:Y_2O_3:MgO:SiO_2$ was used. Samples were then heated at 25°C/min. to 1550°C and immediately cooled at the same rate. The 1550°C temperature was chosen because it was believed to be higher than that necessary to promote reaction. X-ray diffraction showed that the resulting material was alpha-silicon nitride, similar to the UBE E-10 material. This alpha phase is necessary for the liquid phase conversion to the beta phase. It is the beta silicon nitride that is responsible for the self-reinforcing elongated grain morphology.

The UBE E-10 silicon nitride has been shown to have approximately 3% of amorphous silica on the surface.[4] Taking this into account, the total composition ratio of nitride and oxides shifts to 92:2:2:4 $Si_3N_4:Y_3O_3:MgO:SiO_2$, respectively. The oxide fraction in a 1:1:2 ratio was fired to 1450 and 1500°C to determine the minimum temperature required to promote the formation of the ternary oxide glass. Samples fired to both 1450 and 1500 were found to convert to the yttrium and magnesium silicates. Samples fired to 1450°C appeared to be less vitreous than the 1500°C samples.

It was decided that for prereaction, the total batch would be fired to 1425°C. Figures 1 and 2 contrast the appearances of the co-dispersed powders and the preferred jet-milled powders. It can be seen that the co-mixed powders contain loosely held agglomerates, while the prereacted powder consists of silicon nitride/glass composite particles. Their shape was spherical, which was an expected result of size reduction by jet-milling. The size of the prereacted powder appeared to be near 1 μm. This corresponds with X-ray Sedigraph data showing the powder to be an average size of 1.1 μm.

---

[*]   Trost Jet Mill, Garlock, Inc.
[†]   Haake Buchler, Rotovisco RV100

174

Figure 1 Co-dispersed powders with fine particles and loose agglomerates (5000X).

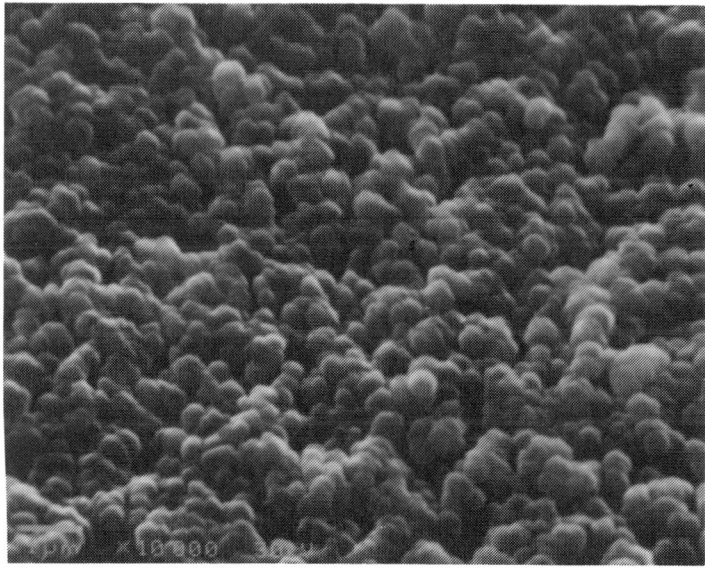

Figure 2 Reacted and jet-milled powder with a larger, more spherical particle morphology due to the formation of cemented agglomerates (10000X).

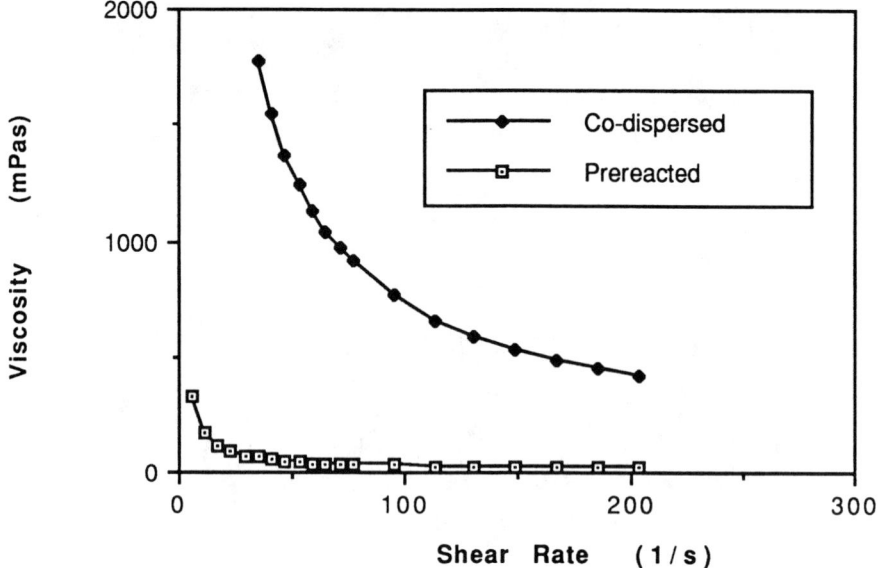

Figure 3 Viscosity vs. shear rate curves demonstrating the reduction of viscosity and pseudoplasticity through the use of prereacted rather than co-dispersed powders.

The relative ease of producing an aqueous dispersion of the co-mixed and pre-reacted powders was determined by examining the deflocculation character of each system. Figure 3 contrasts the viscosity behavior of the co-mixed and prereacted powders. As was expected, it was difficult to obtain a high solids, low viscosity suspension with the co-dispersed powders. For a 34 volume percent suspension, a highly pseudoplastic rheology resulted. The viscosity at a shear rate of 10 s^{-1} was 4050 mPas. It was found that higher solids loading could not be achieved without a very gelled suspension resulting.

When compared to the co-mixed powder, the prereacted powder could be made into a higher solids, lower viscosity suspension. Suspensions containing 38 volume percent solids were prepared with a viscosity as low as 175 mPas at 10 s^{-1}. Figure 3 shows this material to be significantly less pseudoplastic than the co-dispersed powders. This corresponds to work by Cohn, who showed that for prereacted alumina/glass powders a single surface chemistry powder resulted, making stable suspensions more readily prepared than when co-dispersing different powders.

SUMMARY

It was shown that a more stable aqueous dispersion can be produced by prereaction of Si_3N_4, Y_2O_3, MgO, and SiO_2, than by co-dispersion of these same powders. Prereacted powders fired to 1435°C resulted in the formation of a glass that coated the silicon nitride. When cooled and milled, silicon nitride/glass composite particles resulted. A low viscosity (175 mPas) aqueous slip with solids loading of 38 volume percent was achieved with prereacted material, compared with the co-mixed powders whose viscosity at 34 volume percent loading exceeded 4000 mPas.

REFERENCES

1. R. Coe, R. Lumby, M. Pawson, "Some Properties and Applications of Hot-Pressed Silicon Nitride," in Special Ceramics 5, Proceedings of the 4th Symposium on Special Ceramics, Stoke-on-Trent, 1970.
2. S. T. Buljan, J. G. Baldoni, and M. L. Hukabee, "Si_3N_4-SiC Composites," Am. Ceram. Soc. Bull., Vol. 66, No. 2, 1987.
3. A.J. Pyzik and D. R. Beaman, "Processing, Microstructure, and Properties of Self-Reinforced Silicon Nitride," presented at the 92nd Annual Meeting of the American Ceramic Society, 1990.
4. R. deJong,"Incorporation of Additives into Silicon Nitride by Collioday Processing of Metal Organics in an Aqueous Medium," Ph.D. thesis, Rutgers University, 1990.
5. M. Cohn, "A Study of the Sintering Behavior of Alumina and Glass Powders and Prereacted Alumina/Glass Composite Powders," Ph.D. Thesis, Rutgers University, 1991.

SINTERING OF ALUMINA AND ZIRCONIA GREENS OBTAINED VIA SLIP CASTING AND PRESSURE SLIP CASTING

A. Salomoni, I. Stamenkovic, A. Tucci, L. Esposito
Italian Ceramic Center, Via Martelli 26, 40138 Bologna, Italy

ABSTRACT

The casting behavior and microstructures of green and sintered bodies of alumina and zirconia powders were studied. The slips investigated contained from 20 to 43% (v/v) of solids and different deflocculants, and had values of pH ranging from 9 to 10. The green samples were obtained by casting into gypsum molds and by pressure slip casting at 10 MPa. The rates of pressure slip casting, depending on powder characteristics and slip rheology, exceeded the casting rates in plaster molds by one to two orders of magnitude. No substantial differences were found either between green densities or microstructures of the cast or pressure cast samples. The microstructure of green samples depended on characteristics of suspended powder. The surviving agglomerates in the slips considerably deteriorated the homogeneity of the green bodies. Well dispersed powders led to sintered bodies with homogeneous, fined-grained microstructures, while the presence of agglomerates led to sintered bodies containing clusters of pores or large aggregates of grains surrounded by cavities.

INTRODUCTION

The growing interest for developing the microstructures and mechanical strengths of "hot-isopressed-like ceramics,"[1] using conventional densification techniques only, has led to studies of the complex relationships between powder properties, shaping parameters, green/sintered microstructures and parameters of sintered materials. Particular interest was paid to the heterogeneities in green bodies that could create flaws, planar arrays of voids, isolated pores and residual stresses.[2] Perhaps the most common heterogeneity is constituted by agglomerates, particularly hard ones, which have highly altered sinterability.

A number of approaches to avoid the deleterious effects of powder defects or defects in the green ware have been studied. These attempts were directed towards:

- introduction of nonconventional and/or development of existing procedures for the synthesis of ceramic powders[3] aimed at producing the "ideal ceramic powder";[4]

- applying various pretreatments such as grinding, sonication, classification, or use of electrolytes or dispersants;[5,6]

- introduction of appropriate shaping procedures such as cold or hot isostatic pressing, wet shaping, centrifugal casting and electrophoretic deposition.[7-12]

The objective of the research reported in this paper was to investigate the essential parameters of both slip casting and pressure slip casting, as representatives of well established and emerging wet forming techniques. It was expected that these forming techniques would be highly appropriate for the production of defect free oxide ceramics presently under development because they can avoid the procedures of high risk of agglomeration, like spray drying or granulation.

EXPERIMENTAL PROCEDURE AND RESULTS

Two of the oxide powders widely used in our laboratory were studied: high purity alumina and yttria stabilized zirconia. The alumina powder, surface area of 10 m^2/g, consisted of weak, irregular agglomerates ranging up to 100 μm. The zirconia powder, surface area of 7 m^2/g, consisted of fairly hard agglomerates up to 86 μm in diameter.

The slips were prepared in a centrifugal mill equipped with a zirconia jar and balls; the castable slips were obtained after a milling time of 30 minutes. Various deflocculants,[*] having a firing residue declared by producer lower than 0.05 wt %, were used to prepare the slips containing 43% (v/v) of alumina and 33% (v/v) of zirconia. The resulting slips had a pH of 9-10.

Two shaping techniques were employed: slip casting in plaster molds and pressure slip casting. The slip casting was carried out in a high quality gypsum molds whose active capillary volume was approximately 50%. The cast samples were in the form of disks, 28 mm in diameter with thicknesses ranging up to 8 mm. The influence of deflocculant content on casting rates was followed for a broad range of deflocculant content, and the most castable slips were then chosen for all further experiments. The viscosity of the alumina and zirconia slips was measured using a rotational viscometer. The viscosities and casting curves obtained are reported in Fig. 1-2. Scanning electron micrographs of as cast samples are shown in Fig. 3-5.

---

* C—Dolapix CE 64, carbonic acid without alkalies, "Zschimmer-Schwartz" (Germany); P—Dolapix PC 33, synthetic polyelectrolyte without alkalies, "Zschimmer-Schwartz" (Germany); D—Darex 32/A, ammonium salt of carboxylic polyelectrolyte "Grace Italiana" (Italy).

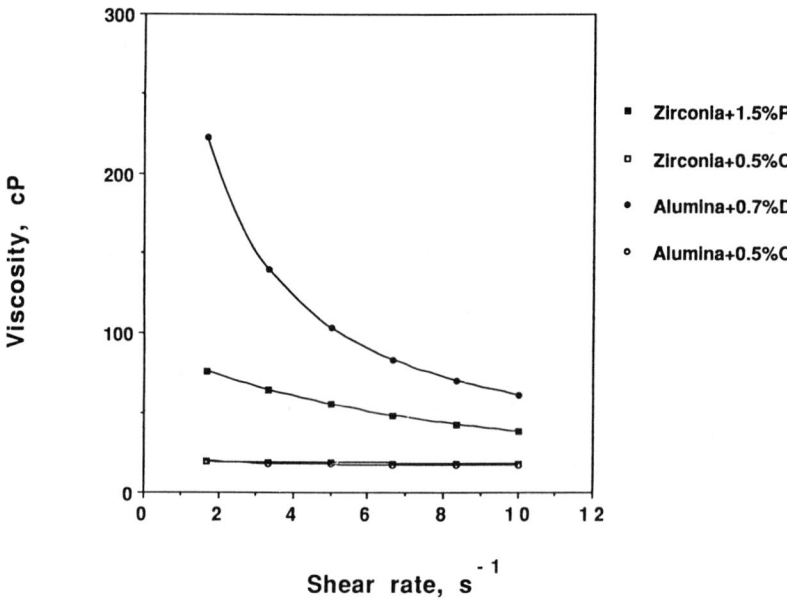

Figure 1 Viscosity of alumina and zirconia slips.

The pressure slip casting was carried out using a hydraulic press expressly designed for this purpose. The slips containing 20% (v/v) of solids and previously chosen deflocculant contents were prepared for the pressure slip casting experiments. A pressure of 10 MPa was used for all the experiments. The samples obtained were in the form of disks, 60 mm in diameter and thicknesses up to 12 mm; the relative casting curves are reported in Fig. 6. The sintering thermal cycle was: heating up to 1500°C at the rate of 60°C/h with two soaks, the first one at 800°C for 1h and the second one at 1500°C for 2h; the cooling rate was 60°C/h.

The densities of the green samples were measured geometrically. The green densities of the cast and pressure cast samples varied within following ranges: zirconia + 1.5% (w/w) P: 46-49% of the theoretical density (TD);
zirconia + 0.5% (w/w) C: 49-53% TD;
alumina + 0.7% (w/w) D: 60-64% TD;
alumina + 0.5% (w/w) C: 61-64% TD.

The water displacement method was used to determine the densities of the sintered samples. The sintered samples obtained by both of the shaping techniques investigated were characterized by densities between 99.2% and 100% TD. Scanning electron microscopy (SEM) was used to analyze the sintered samples; some of the micrographs obtained are shown in Fig. 7-9.

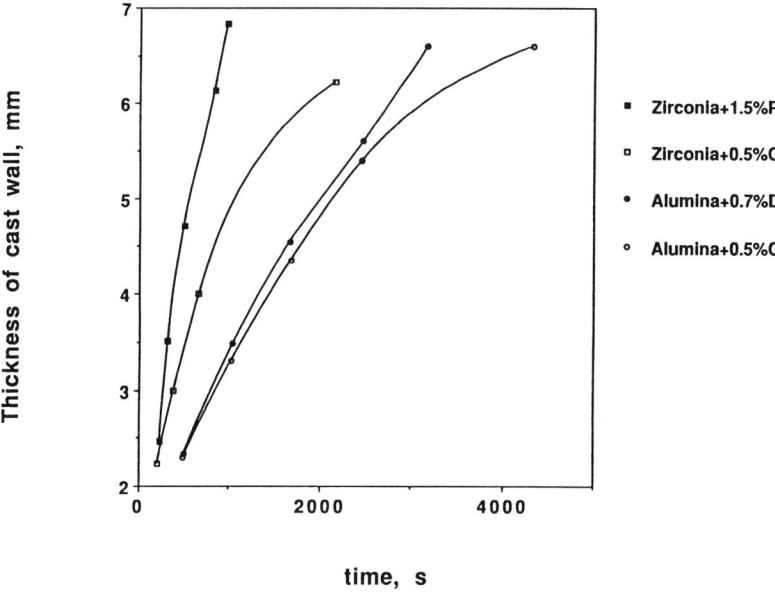

Figure 2 Casting curves of alumina and zirconia slips.

DISCUSSION

Based on the viscosity curves, Fig. 1, one could conclude that the deflocculant C produced the slips behaving as Newtonian liquids. The slips prepared with deflocculants P and D, however, were characterized by higher viscosities and considerable dependence on shear rates. These slips required immediate casting and/or additional mechanical treatments.

The alumina powder investigated was composed of very fine unit particles whose diameters varied within the narrow range of 0.2 to 0.3 μm, and their weak binding forces allowed high levels of deagglomeration to be reached. Consequently, well dispersed alumina particles significantly inhibited the casting kinetics, as can be seen from the casting rates, Fig. 2. Due to the efficient packing of the subagglomerates of these slips, the green bodies produced were characterized by a high green density, 64% of the theoretical density, and high homogeneity, Fig. 3. The starting zirconia powder was characterized by two contrasting parameters; its unit particles, about 0.2 μm, showed a pronounced agglomeration generating relatively consistent, differently sized agglomerates. The low viscosity slips, those containing deflocculant C, permit more intensive deagglomeration of suspended powder. These particles had a higher packing efficiency and led to lower casting rates, Fig. 2, and a more homogeneous microstructure of the green body, Fig. 4. The slip

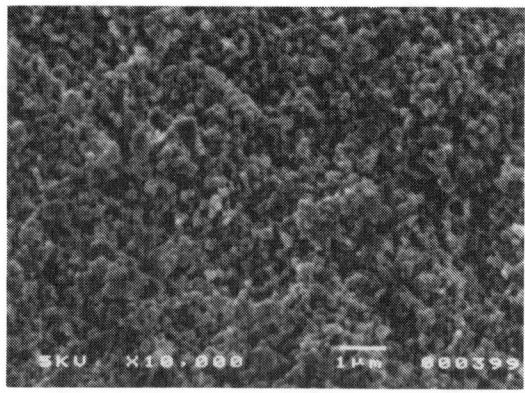

Figure 3 SEM micrograph of cast alumina + D green body.

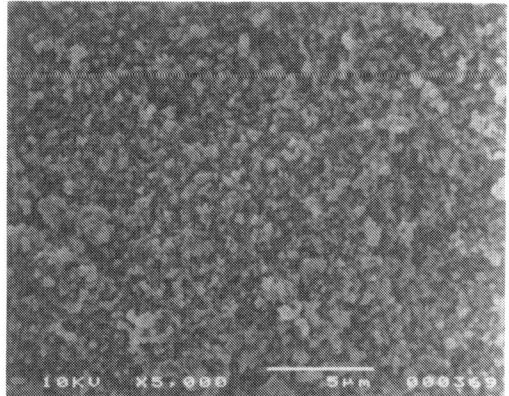

Figure 4 SEM micrograph of cast zirconia + C green body.

containing deflocculant P was characterized by a higher viscosity which slowed-down the process of zirconia deagglomeration and resulted in a more permeable green body as well as higher casting rates, Fig. 2. The green bodies obtained were characterized by imperfections in the form of agglomerates, Fig. 5. Qualitatively the same but more pronounced values of casting rate were obtained with pressure slip casting, Fig. 6. So, the casting rate for pressure slip casting were from one to two orders of magnitude higher as compared with those obtained for slip casting. This once again stresses the growing importance of pressure slip casting in the field of advanced ceramics.

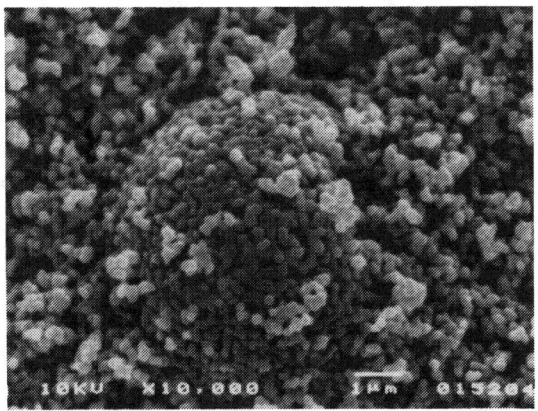

Figure 5 SEM micrograph of cast zirconia + P green body containing agglomerate.

Both of the casting techniques led to the same green densities of the bodies obtained if the same deflocculant and powder was used: up to 64% of TD for the alumina slip and up to 53% of TD for the zirconia slip. Practically no differences were found in the green densities as a function of sample height.

The microscopical examinations showed a similarity between the microstructures of the green bodies obtained by the two casting techniques used. Only the characteristics of the dispersed powders seemed to have a decisive influence on the microstructures of the green bodies. Due to the close packing of constituents obtained by more efficient deagglomeration, the channels created in cast bodies had small diameters which increased the water flow resistance during casting. The microstructures of zirconia bodies were particularly dependent on the level deagglomeration, since the agglomerates in the slip were quite hard. The agglomerates which had survived the contemporaneous actions of milling and deflocculants, in the green bodies were surrounded intimately by subagglomerates, Fig. 5.

All the samples investigated, sintered at 1500° for 2h, reached high levels of density, ranging between 99.2% and 100% of TD, in accordance with the declared sinterability of the powders. The qualitative interdependence between the microstructures of the green and sintered bodies was found. The differences in sinterability of the agglomerates and ceramic matrix caused the elliptical cavities, while the smaller agglomerates were transformed into relatively large, up to 9 μm, clusters of pores, as can be seen in Fig. 7 showing the SEM micrograph of sintered zirconia +P body. The well dispersed powders led to theoretically dense bodies with highly homogeneous microstructures, as can be seen in the SEM micrographs of sintered alumina

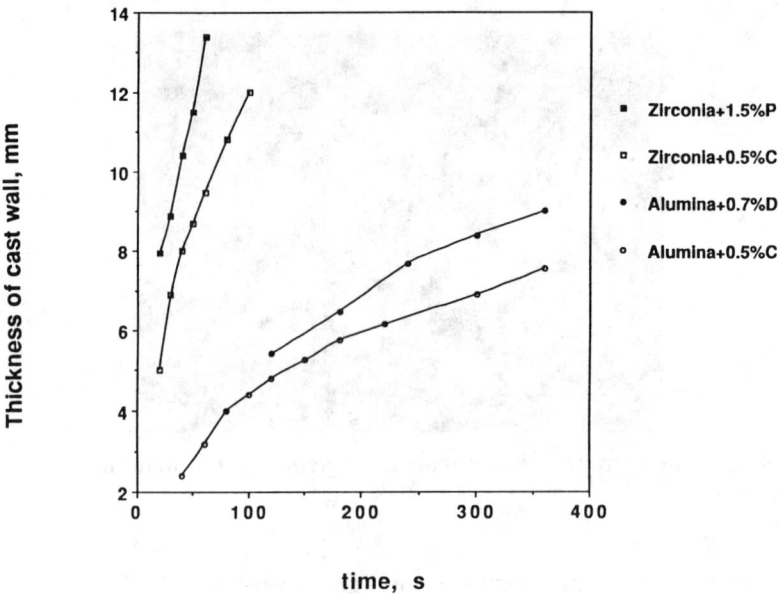

Figure 6 Pressure casting curves of alumina and zirconia slips.

Figure 7 SEM micrograph of sintered zirconia +P body containing cluster of pores.

+ D and zirconia + C bodies showed in Fig. 8 and 9 respectively. A closer, quantitative analysis of the microstructures is in progress.

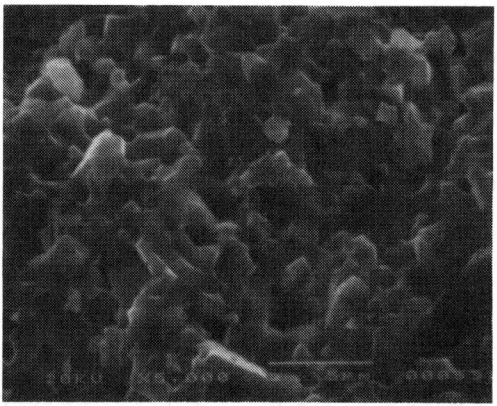

Figure 8 SEM micrograph of sintered alumina +D body.

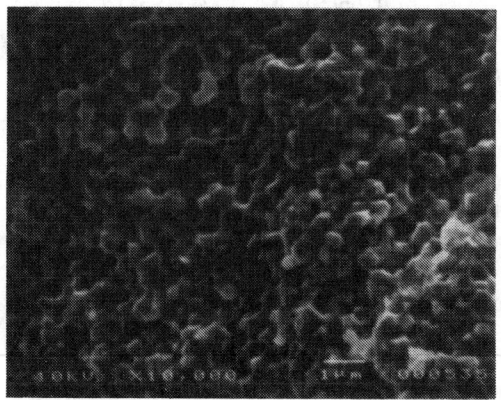

Figure 9 SEM micrograph of sintered zirconia +C body.

CONCLUSIONS

The casting rates of alumina and zirconia powders were prevalently influenced by the powder characteristics and the slip rheology. The casting rates in gypsum molds were from one to two orders of magnitude lower than those of pressure slip casting. The densities and microstructures of the green bodies obtained by slip casting and pressure slip casting were very similar. The qualitative interdependence between the microstructures of the green and sintered bodies was confirmed.

ACKNOWLEDGMENTS

The authors wish to thank S. Degli Esposti and D. Naldi for their help in the experimental work.

REFERENCES

1. J. Sung, P.S. Nicholson, "Strength Improvement of Yttria-Partially-Stabilized Zirconia by Flaw Elimination" J. Am. Ceram. Soc. 71 [9] 788-795 (1988).
2. C.H. Hsueh, "Sintering Behaviour of Powder Compacts with Multihetero-geneities" J. Mat. Sci. 21, 2067-2072 (1986).
3. D.W. Johnson, "Sol-Gel processing of ceramics and glass" Am. Ceram. Soc. Bull. 64 [12]1597-1602 (1985).
4. R.W. Davidge, "Defects in Ceramics-The Targets for NDT" Br. Trans. J. 88, 113-116 (1989).
5. A. Roosen, H. Hausner, Adv. Ceram. Mat. 3[2]131-137 (1988).
6. F.F. Lange, "Powder Processing Science and Technology for Increased Reliability" J. Am. Ceram. Soc. 72[1]3-15 (1989).
7. F.F. Lange, K.T. Miller, "Pressure Filtration Consolidation Kinetics and Mechanics" Am. Ceram. Soc. Bull. 66[10]1498-1504 (1987).
8. F. Harbach, R. Neeff, H. Nienburg, L. Neiler, "Reliable Ceramic Components from Colloidal Suspensions" Cfi-DKG 67[4]130-135 (1990).
9. A. Roosen, H.K. Bowen, "Influence of Various Consolidation Techniques on the Green Microstructures and Sintering Behaviour of Alumina Powders" J. Am. Ceram. Soc. 71[11]970-977 (1988).
10. A. Salomoni, C. Palmonari, I. Stamenkovic, "Pressure Slip Casting of Highly Sinterable Oxide Powders" Am. Ceram. Soc. 92nd Annual Meeting, Poster Session II (1990).
11. F.F. Lange, M. Metcalf, "Processing-Related Fracture Origins: II Agglomerate Motion and Cracklike Internal Surfaces Caused by Differential Sintering" J. Am. Ceram. Soc. 66[6]398-406 (1983).
12. S. Inada, T. Kimura, T. Yamaguchi, "Effect of Green Compact Structure on the Sintering of Alumina Cer. Int. 16,369-373 (1990).

186

KINETICS AND MECHANICS OF CONSTANT PRESSURE FILTRATION OF COLLOIDAL CERAMIC DISPERSIONS

Bhaskar V. Velamakanni[*] and Fred F. Lange
Materials Department, University of California, Santa Barbara, CA 93106

EXTENDED ABSTRACT

The kinetics of constant pressure filtration of coagulated (i.e. weakly attractive) ceramic dispersions was investigated and the filtration results are compared to that of dispersed and flocced dispersions. Ceramic dispersions can be weakly coagulated in the presence of short-range repulsive hydration layers by adding hydrolyzable ions to a dispersed slurry. Such slurries exhibit very high viscosities and yet, be packed to a high density during pressure filtration by lubrication-assisted particle rearrangement due to the short range repulsive potential.[1] In the coagulated state particles sit in a shallow "hydration" minimum (depth between 0 and 50 kT), which is distinctly different from flocced state where particles sit in deep "primary" minimum ($\approx$150 kT). The strength of the coagulated network (and viscosity) can be changed by controlling the ionic strength of the dispersion, hence it presents an ideal to tool for monitoring the kinetics of filtration between dispersed and flocced states.

The kinetics of filtration of 20 vol.% alumina slurries (0.2 μm media particle diameter) was analyzed as per the procedure described by Lange and Miller.[2] The filtration kinetics of all the slurries investigated followed Darcy's law with the assumed condition that particle density exhibited a step function traversing the consolidated body into the slurry, viz., without gradients. The effect of pH and ionic strength (I) on the filtration kinetics is presented in Table 1 as in the form of packing density (φ), permeability (κ) and filtration rate (K) of different slurries. Since permeability of the consolidated layer during filtration increases with decrease in

Editor's Note: This extended abstract was not peer reviewed.

---

* Now at 3M Corporate Research Laboratory, St. Paul, MN 55144.

Table 1 Influence of pH and Ionic Strength on the Kinetics of Constant Pressure Filtration (2.8 mPa).

Slurry State	pH	I (M)	φ	$\kappa\,(m^2 \times 10^{-17})$	K (μm/sec)
dispersed	4.0	0	0.569	2.22	6
coagulated	4.0	0.1	0.571	2.479	7
	4.0	0.5	0.555	3.631	9
	4.0	1.0	0.564	3.098	10
	4.0	1.5	0.562	3.707	12
	4.0	2.0	0.595	4.00	12
flocced	9.0	0	0.492	10.396	24

packing density, the filtration rate is the lowest for dispersed slurries (highest packing density) and the highest for flocced slurries (lowest packing density). For the same maximum packing density, the filtration rate (or consolidation rate) of coagulated slurries increases up to twice that of dispersed slurries with ionic strength. It was shown in this study, when coagulated slurries are filtered, the solids concentration in the slurry column over the consolidated layer gradually decreases from a maximum concentration (that of the consolidated layer) to a minimum concentration (that of the original slurry). In other words, during filtration, a gradient in particle density exists in the consolidating layer that produces a higher permeability, and thus a faster filtration rate. Since no significant difference in maximum particle packing density was observed for different salt additions, it is concluded that the effect of applied pressure on packing density occurs at very low pressures.

REFERENCES

1. B. V. Velamakanni, J. C. Chang, F. F. Lange and D. S. Pearson, "New Method for Efficient Colloidal Particle Packing via Modulation of Repulsive Lubricating Hydration Forces", *Langmuir*, 6, 1323-1325 (1990).
2. F. F. Lange and K. T. Miller, "Pressure Filtration: Kinetics and Mechanics," *J. Am. Ceram. Soc.*, 66 [10] 1498-504 (1987).

NEW PROCESSING METHOD FOR NEAR NET-SHAPED, COMPLEX-SHAPED STRUCTURAL CERAMICS

Bhaskar V. Velamakanni[*] and Fred F. Lange
Materials Department, University of California, Santa Barbara, CA 93106

EXTENDED ABSTRACT

A method has been developed to form complex shaped, near-net-shaped ceramics using a technique which involves vibrating a liquid saturated powder compact, previously consolidated by either pressure filtration or centrifugation, into a complex shaped mold cavity. During vibration, the powder compact becomes sufficiently fluid to fill the cavity, whereas after the vibration is stopped, the powder compact, which now has the shape of the cavity, becomes sufficiently stiff to retain its shape without distortion. After the ceramic is dried in the mold, it was removed from the mold and heat treated to densify. The unique rheology of the powder compact is produced by a new method of manipulating interparticle forces to develop a weakly attractive particle network in the presence of short-range repulsive hydration layers.[1] The water saturated powder compact is produced with the following sequence of operations: 1) forming a dispersed slurry by mixing a powder (≤ 30 volume %) with water at a pH that produces a net surface charge and a highly repulsive interparticle force, 2) adding a sufficient amount of salt to the dispersed slurry to cause particles within the slurry to attract one another, and 3) increasing the volume fraction of particles by either pressure filtration or centrifugation to form a water saturated powder compact with a uniform and very high particle packing density. After consolidation, the attractive forces between particles within the powder compact, saturated with water, are sufficiently strong to prevent body flow, but sufficiently weak to produce flow when the body is subjected to a modest vibration. The saturated powder compacts, when vibrated, exhibit extensive shear thinning which allows the compact to appear sufficiently liquid-like

Editor's Note: This extended abstract was not peer reviewed.

---

* Now at 3M Corporate Research Laboratory, St. Paul, MN 55144.

(plastic flow)—achieved without using polymeric binders and/or plasticizers—through apparent interparticle lubrication.

Since colloidal powder treatments (sedimentation and/or filtration) can be used to eliminate many heterogeneities common to powders and ensure more uniform phase distributions, the present technique can produce ceramics with enhanced reliability. In addition, this new method is not only extremely simple, inexpensive and safe but it also avoids the use of organic solvents, polymeric binders and plasticizers.

REFERENCES

1. B. V. Velamakanni, J. C. Chang, F. F. Lange and D. S. Pearson, "New Method for Efficient Colloidal Particle Packing via Modulation of Repulsive Lubricating Hydration Forces", *Langmuir*, 6, 1323-1325 (1990).

CELLULOSE ETHERS IN TAPE CASTING FORMULATION

K. E. Burnfield and B. C. Peterson
Aqualon Company, Wilmington, DE 19850-5417

ABSTRACT

Ceramic tapes were cast using a hydroxyethylcellulose binder. The green tapes exhibit excellent flexibility and strength. By varying concentrations of the binder and other ingredients during processing, properties of the green tape change. Flexibility, strength, density and binder burnout are examined.

INTRODUCTION

Tape casting techniques for the preparation of ceramic electronic substrates have been well documented.[1-6] Typical systems include the use of organic solvents and polyvinyl butyral[3] or water and latex binders. Other examples of tape casting systems include nitrocellulose and organic solvents[4] or other cellulose ethers and water.[5] The process allows for a thin sheet of ceramic to be prepared. The ceramic sheet can be cut or stamped to produce a near net shape ceramic piece when fired. The green tape properties must be optimized for easy casting and processing.[6] These properties include green strength, flexibility, density, thickness and smoothness.

EXPERIMENTAL PROCEDURE

Ceramic green tapes were prepared according to the recipe in Table 1 using hydroxyethylcellulose (HEC). All ingredients were added to a ball mill containing ceramic grinding media and ball milled for 6 hours. The slips were poured onto a nonadhering polyester sheet in front of a doctor blade adjusted to 0.127 or 0.254 cm. The blade was pulled forward producing a wet tape. The tape was allowed to dry under ambient conditions 16 hours before removal from the sheet. Dry tape thickness was measured using a micrometer.

Tensile testing was conducted on an Instron Corporation series IX automated system. Testing was done at 50% relative humidity and 25 degrees C. Crosshead speed was set at 2.54 cm/min. Sample size was 1.28 X 5.12 cm. The instrument

191

Table 1 Tape Casting Formulation.

	Wt%
Mineral (alumina)	34-50%
Binder (Natrosol® 250LR, HEC)	3-7%
Plasticizer (PEG)	0.8-4.5%
Dispersant (Darvan® 821A)	0.2-1.5%
Water	41-62%

computer calculations for tensile strength were adjusted for variable tape thickness.

Density was determined using displacement techniques. A weighed quantity of green tape was submerged in a non-solvent contained in a graduated cylinder. The increase in volume measured by the level of the non-solvent in the graduated cylinder was used to calculate green tape density.

Binder burnout testing was accomplished by standard thermal gravimetric analysis techniques. The binder was run neat and in the presence of alumina powder. The binder-alumina sample was prepared by making a slip composed of 5% HEC binder, 45% alumina and 50% water. The slip was oven dried at 50 degrees C to remove the water. The binder and the binder-alumina sample were tested by TGA using a ramp rate of 20 degrees C/minute in the presence of air.

RESULTS AND DISCUSSION

All testing was done on unfired ceramic tape. Alumina tapes cast at 0.127 and 0.254 cm have equal shrinkages of approximately 77% in the z direction (tape density = 2.6g/cc). Shrinkages in the x and y directions are less than 1% and 3%, respectively. Figure 1 shows the dependence of green tape strength on binder concentration. Tape strength increases with increasing binder content. Higher degrees of increase are seen above 4% binder. Also, Figure 1 illustrates data generated by substituting a higher molecular HEC [Natrosol 250MR (MW=300,000) for Natrosol 250LR (MW= 90,000)]. Increasing the molecular weight of the binder significantly increases the strength of the tape. Increasing the concentration of binder decreases the green tape density as shown in Figure 2. A linear relationship exists between binder content and density.

Contributions of the dispersant and plasticizer to strength and flexibility are shown in Figure 3. For the purposes of this paper, their individual contributions are not isolated. Instead, a constant ratio of dispersant to plasticizer is maintained. Figure 3 shows a linear relationship of decreasing strength with increased dispersant:plasticizer concentration. Conversely, elongation or strain relating to flexibility

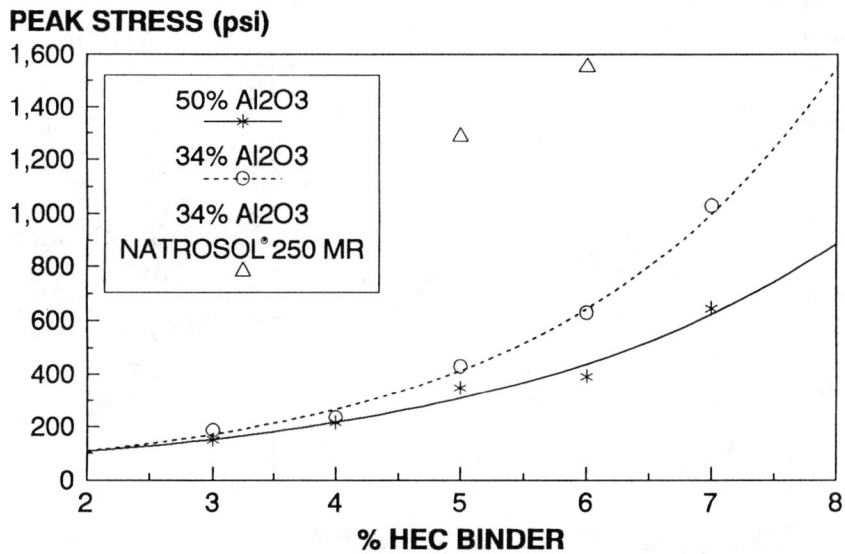

Figure 1 Ceramic tape strength testing 0.88% plasticizer, 0.25% dispersant.

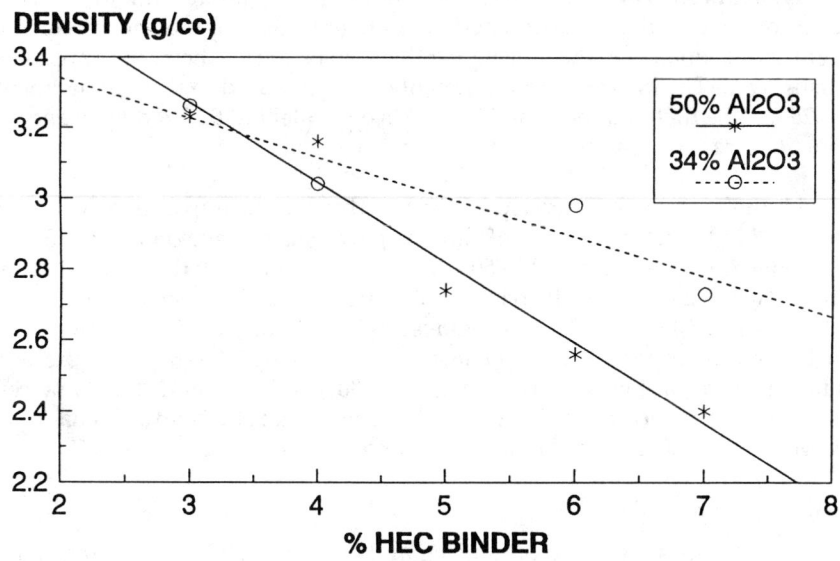

Figure 2 Ceramic tape strength testing 0.88% plasticizer, 0.25% dispersant.

193

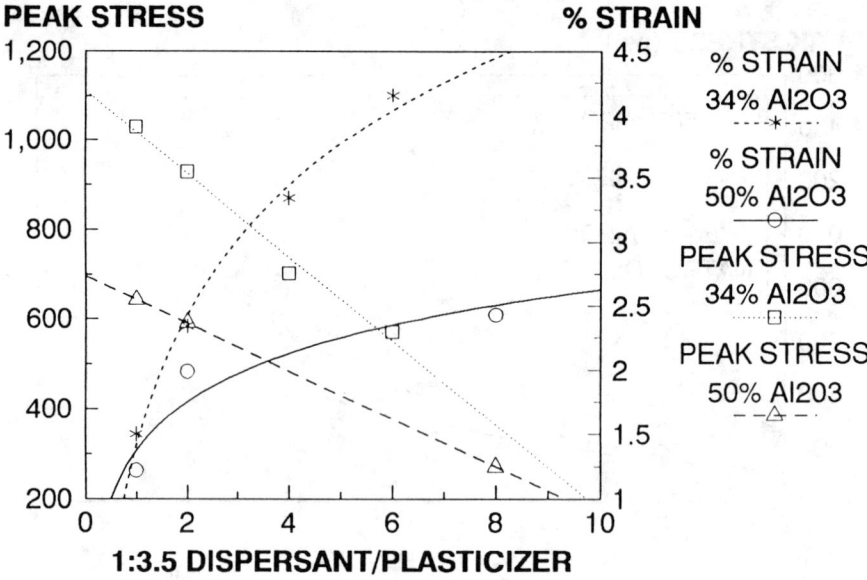

Figure 3 Ceramic tape tensile testing 7% HEC binder.

increases significantly with dispersant:plasticizer concentration. Although no evidence is offered in this paper, experience suggests that significant changes in strength and flexibility are more heavily influenced by the plasticizer as opposed to the dispersant. The lowest concentration of dispersant used in these experiments was 0.25%. The first data point in Figure 3 is equivalent to 0.25% dispersant and 0.875% plasticizer (1:3.5).

Figure 4 is the TGA curve generated for the HEC binder. A comparative TGA curve of the HEC binder in the presence of alumina powder is presented in Figure 5. The binder alone degrades from 230 to 500 degrees C, with 75% of the binder decomposing between 280 to 390 degrees C. The TGA of HEC alone shows 3 major isotherms. Only 40% of the HEC is removed below 310 degrees C. Fifteen percent of the binder decomposes above 390 degrees C leaving 0.26% ash. In the presence of alumina the major stage of burnout occurs at 230 to 310 degrees C. Eighty percent of the binder is removed below 310 degrees C. The TGA of HEC and alumina shows only two major isotherms. Complete burnout occurs before 425 degrees C.

CONCLUSIONS

Green ceramic tapes were prepared using an HEC binder. Tensile testing shows that green strength increases with the HEC binder content and molecular weight. Conversely, density decreases with increasing binder content. Green strength de-

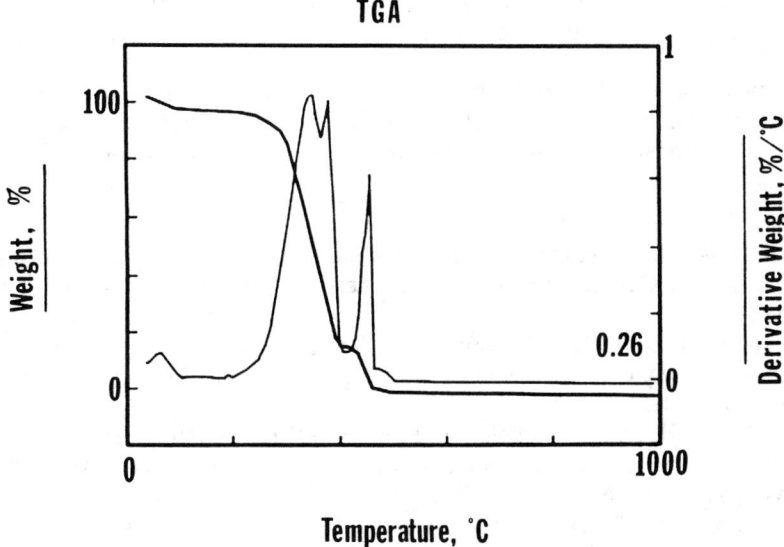

Figure 4 TGA of the HEC binder.

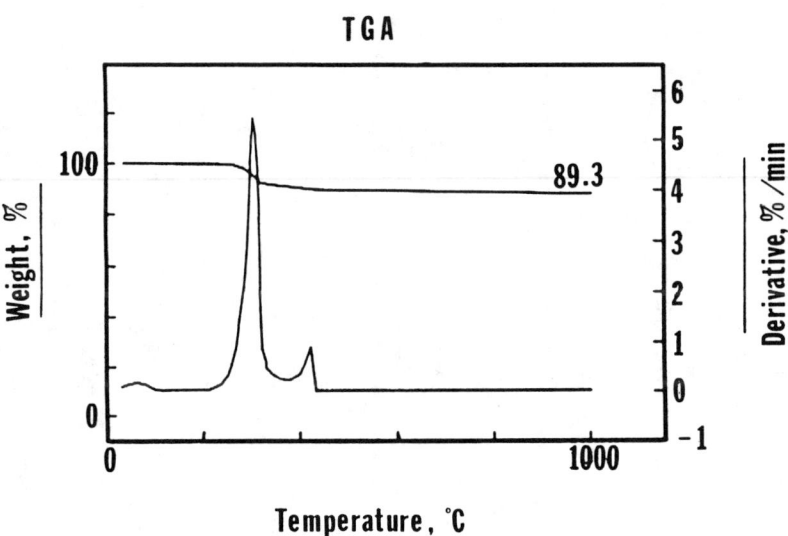

Figure 5 TGA of the HEC binder and alumina.

195

creases with the dispersant:plasticizer content, but elongation or flexibility increases with the dispersant:plasticizer content.

TGA studies of the HEC binder in the presence of alumina prove that the binder burns out differently from the polymer alone. In both experiments the polymer burns out completely, but the rate and temperature of burnout differ.

REFERENCES

1. Karas, A., T. Kumagai and W. R. Cannon, "Casting Behavior and Tensile Strength of Cast BaTiO3 Tape," Advanced Ceramic Materials, 3 [4] 347-77 (1988).
2. Shanefield, D. J., "Tape Casting for Forming Advanced Ceramic," pp 4855-8, Encyclopedia of Material Science and Engineering, ed. M.B. Bever (Pergamon: New York, 1985).
3. Mistler, R. E., D. J. Shanefield and R. B. Runk, "Tape Casting of Ceramics," pp 411-48, Ceramic Processing before Firing, ed. G. Y. Onoda and L. L. Hench (John Wiley and Sons: New York, 1978).
4. Thompson, J. J., "Forming Thin Ceramics," Ceramic Bulletin, 43 [9] 480-1 (1963).
5. Burnfield, K. E. and D. Schweizer, "The Effects of Organic Additives on Ceramic Body Processing," Advanced Ceramics 90, SME Technical Paper EM90-136 (1990).
6. Hyatt, E. P., "Continuous Tape Casting for Small Volumes," Ceramic Bulletin, 68 [4] 869-70 (1989).

THE EFFECTS OF SOLVENTS AND BINDERS ON THE PROPERTIES OF TAPE CASTING SLURRIES AND GREEN TAPES

Jane-Chyi Lin, Tsung-Shou Yeh, Chian-Lii Cherng, and Chien-Min Wang
Materials Research Laboratories, Industrial Technology Research Institute,
Chutung, Hsinchu, 31015, Taiwan, R.O.C.

ABSTRACT

A ceramic-glass system containing alumina, forsterite, and borosilicate glass that has low dielectric constant and can be sintered at low temperatures (850-1000 °C) was investigated. Slurries composed of solvents, dispersants, binders, plasticizers, and the ceramic-glass powders were made to flat sheets by tape casting method. The various organic solvents (MEK, toluene, ethanol) and binders (PVB) had significant influence on the properties of the slurries and green tapes. The viscosity of the slurries decreased as the polarity of the solvent increased. Lower viscosity and less shear-thinning slurries were obtained with binary solvent systems. The slurry prepared with MEK/toluene mixtures with high hydroxyl content PVB showed shear thinning behavior and the green tape had flat and smooth surface. Shear thickening was observed for slurries prepared with MEK/toluene mixtures with low hydroxyl content PVB that resulted in green tapes with cracks and wavy surfaces. Shear-thickening suspensions, however, were better dispersed than those of shear-thinning ones. Higher green density and finer pore size were obtained for cast tapes prepared from shear-thickening suspensions.

INTRODUCTION

In the past, alumina had been the major material for making ceramic substrates and packages in the electronic industry.[1] Recently, ceramic-glass materials were proposed as potential replacements for alumina because of their lower dielectric constants (4-8) and lower sintering temperatures (850-1000 °C).[1,2] Many studies have been carried out to investigate the production of ceramic-glass substrates which, in majority, were made of thin ceramic sheets from tape casting method.[3] In general, the casting slurry is a complex system containing solvents, binders, plasticizers, dispersants, and ceramic powders. Each component has substantial effect on the slurry flow characteristics which in turn affect the microstructure and

197

densification behavior of the green sheet. In order to improve product quality, the component-slurry-green tape relationships need to be understood. In this paper, the effects of solvents and binders on the rheological behaviors of slurries and the properties of green tapes were studied.

EXPERIMENTAL

The ceramic material used in the present study was a mixture of borosilicate glass,[*] alumina,[†] and forsterite[*] powders. Forsterite and glass powders were first wet ball-milled to desired particle size distribution with a median size range of 2-3 μm. Two binary solvent systems, i.e., methylethyl ketone (MEK)/toluene and MEK/ethanol, were employed. All slurries were prepared with 30 vol% solids loading containing various ratios of the two solvents. Two polyvinyl butyral (PVB)[‡] binders with similar molecular weight were designated as PVB-H (high hydroxyl content) and PVB-L (low hydroxyl content). PVB-L was employed exclusively in the study of the solvent effects and both were used for comparison in the investigation of the binder effects. Dibutyl amine and dibutyl phthalate were the dispersant and plasticizer, respectively. The formulations of the slurries were as follows:

powder	100 g	solvent	60-70 g
dispersant	0-3 g	binder	4-10 g
plasticizer	2-4 g		

The mixtures were ball-milled for 20 h to produce homogeneous slurries. The rheological characteristics of the slurries were determined by a concentric cylinder viscometer.[§] After the viscosity measurement, a rotary evaporator was used to remove air bubbles and excess solvents. Doctor-blade caster was then used to prepare thin green tapes on Mylar films. The green density and pore size distribution of the tapes were measured by a mercury porosimeter.[¶]

RESULTS AND DISCUSSION

Effects of Solvents

The major function of a solvent is to act as a dispersing vehicle so that a castable slurry can be obtained. In general, the selected binder, plasticizer, and dispersant should be soluble in the chosen solvents which should be inert to the ceramic powders. Borosilicate glass used in this study was found to react with water and, therefore, organic solvents were used. Figure 1 shows the relative viscosity at various shear rates for the slurries made of different solvents. Relative viscosity was obtained by dividing the slurry viscosity by the vehicle (i.e., solvent + binder

---

[*] Prepared by Materials Research Laboratories, ITRI, Taiwan, R.O.C.
[†] AL-160SG-4, Showa Denko Co., Tokyo, Japan.
[‡] Butvar, Monsanto Co., St. Louis, MI
[§] RV-20, Haake Mess-Technik GmbH u. Co., Federal Republic of Germany.
[¶] Autopore II 9220, Micromeritics Instrument Corp., Norcross, GA.

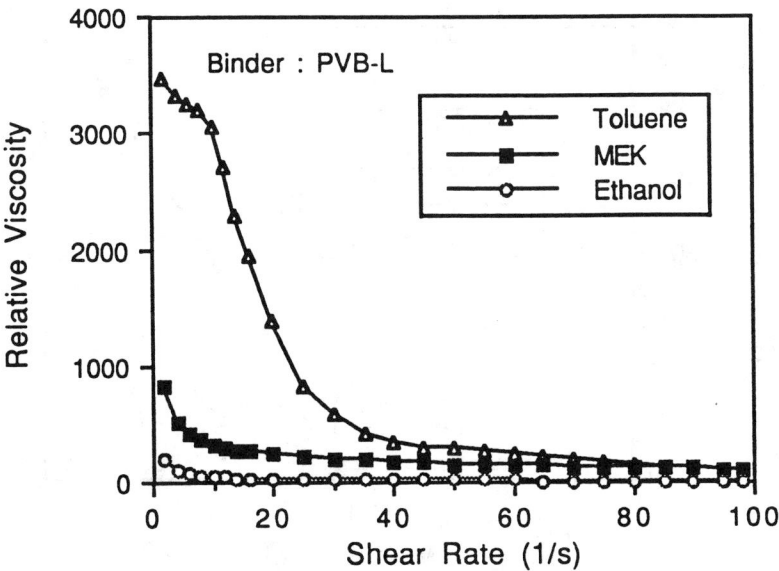

Figure 1 Relative viscosity vs. shear rate for slurries prepared with various solvents.

+ plasticizer + dispersant) viscosity. Only slight shear-thinning was observed for the slurry prepared with ethanol which has the highest polarity among the solvents used. This slurry also had the lowest relative viscosity, indicating better dispersion. As the polarity of the solvent decreased, the degree of shear-thinning increased.

Due to the complexity of the tape cast system, it is common to use a combination of solvents to increase the compatibility of solvents and other slurry constituents. The solvent ratio in a binary solvent system may greatly affect the behavior of the slurry. Figure 2 shows the relative viscosity of the slurries prepared with varied MEK/ethanol ratios. Both single-solvent slurries showed shear-thinning behavior. This pseudoplasticity was reduced when both solvents were present and MEK/ethanol ratio of 50/50 gave the lowest slurry viscosity. This indicates that better dispersion can be obtained with a mixture of the solvents. In a study conducted by Sacks and Scheiffele on the effect of solvents (methanol and methyl isobutyl ketone) upon the properties of alumina suspensions, they also reached similar results.[4] Similar trend was also found for slurries prepared with MEK and toluene. It should be noted that slight shear thickening (i.e., viscosity increases as the shear rate increases) was obtained at low shear rate range for the slurries prepared with the MEK/toluene ratios of 25/75 and 50/50.

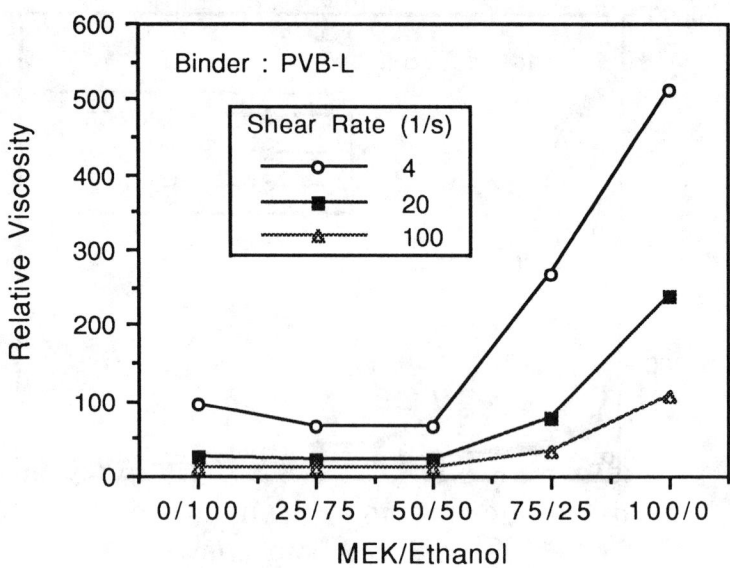

Figure 2 Relative viscosity at three chosen shear rates for slurries prepared with various MEK/ethanol ratios.

Plot of relative viscosity vs. shear rate for the slurry with MEK/toluene = 50/50 is shown in Figure 3. The viscosity of the slurry with MEK/ethanol = 50/50 is also given for comparison. The MEK/ethanol slurry was highly shear-thinning. The MEK/toluene slurry was shear thickening at low shear rates up to 20 1/s above which the viscosity decreased as shear rate increased. In general, shear thickening occurs only with percolation structures.[5] For a particulate system, percolation does not occur until the solids loading reaches a critical value which is normally greater than 50 vol%.[5] However, the solids loading in the present study was only about 30 vol% which is far below the normal critical value. One possible explanation of the shear-thickening behavior is that the slurry was well dispersed at rest which gave low viscosity and, when shear was applied, the ceramic powders in the slurry formed aggregated structures which caused the increase in viscosity.[6] As shear rate increased further, viscosity decreased due to the breakdown of the weak aggregated structures. In the present system, shear thickening is also related to the PVB hydroxyl content. This is discussed later in the text.

Slurries with shear-thinning behavior are usually more desirable for tape casting operation. During casting process, slurry viscosity decreases under the stress generated by the doctor-blade. Immediately after the stress is released, the slurry viscosity returns to high level. This restricts the flow of the suspension and thus

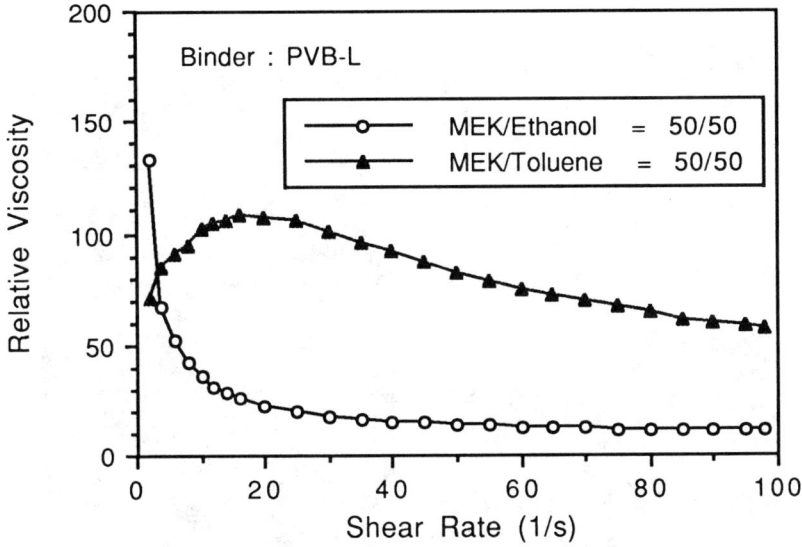

Figure 3 Relative viscosity vs. shear rate for slurries prepared with either MEK/toluene = 50/50 or MEK/ethanol = 50/50.

enables accurate control of the tape dimensions. A shear-thickening slurry behaves just the opposite when it is subjected to casting operation. When the slurry leaves the blade, reduction of the viscosity causes the flow of the suspension and difficulties in dimension control. Figure 4 shows top surface photograph of green tapes prepared with (a) MEK/ethanol = 50/50, and (b) MEK/toluene = 50/50. The MEK/ethanol green tape had a smooth flat top surface. On the other hand, the MEK/toluene green tape had a very rough surface. The latter tape also exhibited severe cracking during drying process. This suggests that the slurry shear thickening was related to the inhomogeneous drying and non-uniform shrinkage of the green tape. Figure 4 also shows the SEM micrographs of green tapes prepared with (c) MEK/ethanol = 50/50, and (d) MEK/toluene = 50/50. The microstructure of the MEK/toluene tape was more uniform and had better packing. This is also reflected in the pore size data obtained by mercury porosimetry (Figure 5). The porosity of MEK/toluene tape after binder burnout was lower (45%) than that of the MEK/ethanol tape (56%). The medium pore size of the MEK/toluene tape was also smaller (0.36 vs. 0.59 μm). These results are the evidence that the shear-thickening slurry was better dispersed at rest.

Effects of Binders

The major function of the polymeric binders is to provide the green tape with proper

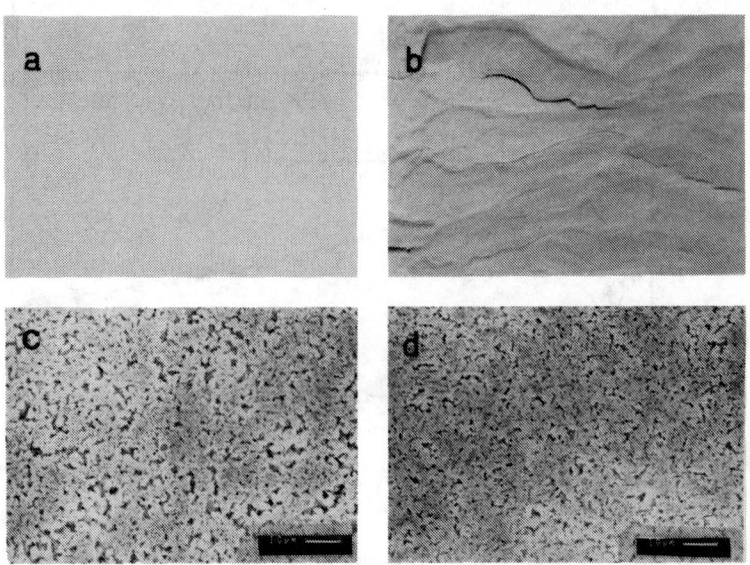

Figure 4 Top surface photograph of green tapes prepared with (a) MEK/ethanol = 50/50, and (b) MEK/toluene = 50/50. SEM micrograph of green tapes prepared with (c) MEK/ethanol = 50/50, and (d) MEK/toluene = 50/50.

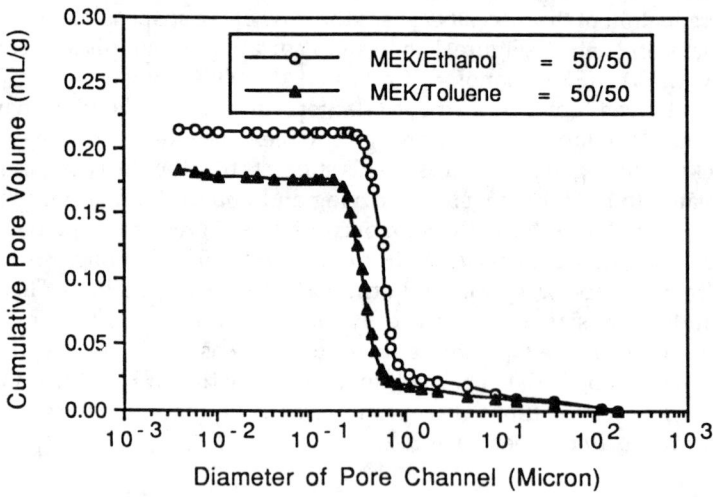

Figure 5 Plot of cumulative pore volume vs. diameter of pore channel for green tapes from (a) MEK/ethanol = 50/50, and (b) MEK/toluene = 50/50.

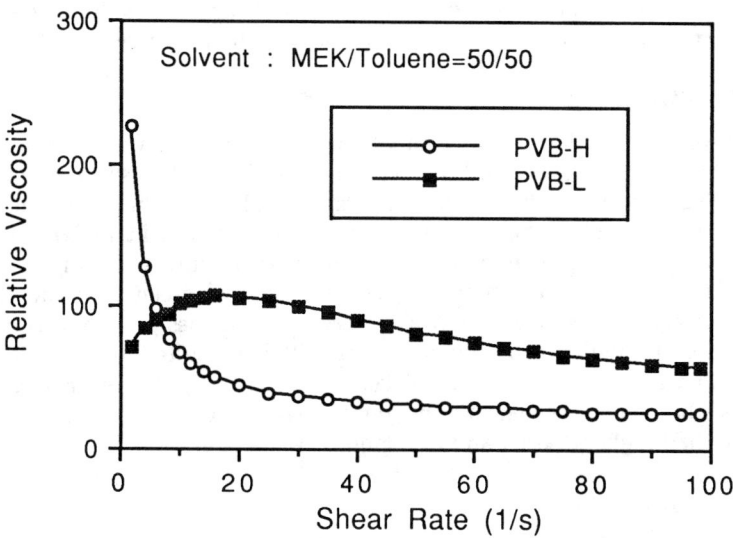

Figure 6 Relative viscosity vs. shear rate for slurries prepared with MEK/toluene = 50/50 and various PVB binders.

strength and flexibility for further handling. In addition, binders can also be used as dispersants to modify the rheological characteristics of the slurry that consequently affect the microstructures of the green tapes.[4] As mentioned before, the slurry containing PVB-L binder and MEK/toluene = 50/50 solvent showed shear thickening at low shear rates. However, when PVB-L was replaced by PVB-H which has a higher hydroxyl content, the slurry rheology changed from shear thickening to pseudoplastic. The two relative viscosity curves are shown in Figure 6. The slurry prepared with PVB-H was more viscous at rest and highly shear-thinning once sheared. Although the cause of the difference in the flow properties is not yet fully understood, it is likely to be related to the PVB hydroxyl content. The hydroxyl groups in the PVB molecule tend to form hydrogen bonding with other PVB molecules and the powder surface. Therefore, an increase in the hydroxyl content may enhance PVB adsorption. This can affect the conformation of the adsorbed binder or the distribution of the binder molecules in the suspension. Consequently, the aggregational behavior of the binder and solid particles may be changed. This may have caused the disappearance of the shear-thickening behavior observed in the PVB-L slurry. Possible mechanism of the shear thickening was mentioned earlier. The change of flow characteristics of the PVB-L slurry implies that the suspension changed from a well dispersed state to form a percolation structure due to shear. This structure was weak and the flow was quickly dominated by the shear-thinning effect. The data suggest that the binder played an important role in the slurry dispersion and, therefore, the slurry rheology. A detailed investigation will be carried out to examine the effect of binder upon the slurry shear thickening.

CONCLUSIONS

Organic solvents (MEK, toluene, ethanol) and PVB binders have significant influence on the properties of the tape casting slurries and green tapes in a ceramic-glass system. The viscosity of slurries decreased as the polarity of the solvent increased. Lower viscosity and less shear-thinning slurries were obtained with binary solvent systems. The slurry prepared with MEK/toluene mixtures with high hydroxyl content PVB showed shear thinning behavior and the green tape had flat and smooth surface. Shear thickening was observed for slurries prepared with MEK/toluene mixtures with low hydroxyl content PVB that resulted in green tapes with cracks and wavy surfaces. Shear-thickening suspensions, however, were better dispersed than those of shear-thinning ones. Green tapes cast from shear-thickening slurry had higher density and finer pore size.

ACKNOWLEDGMENTS

This work was supported by the Ministry of Economic Affairs under Contract No. 34N1300 through Industry Technology Research Institute. The authors are very grateful to Dr. A. C. Young for his suggestions.

REFERENCES

1. R.R. Tummala, Ceramic Packaging, pp455 in Microelectronics Packaging Handbook, Edited by R.R. Tummala and E.J. Rymaszewski, Van Nostrand Reinhold New York, 1989.
2. M. Takabatake, J. Chiba, and Y. Kokubu, Composition for Multilayer Printed Wiring Board, U. S. patent, No. 4593006, June 3, 1986.
3. J. C. Williams, Doctor-Blade Process, pp173, Ceramic Fabrication Processes, Vol 9, Edited by Franklin F.Y. Wang, Academic Press Inc., New York, 1976.
4. M.D. Sacks and G.W. Scheiffele, Polymer adsorption and Particulate Dispersion in Nonaqueous Al_2O_3 Suspensions Containing Poly(vinyl Butyral) Resins, pp175 in Multilayer Ceramic Devices, American Ceramic Society, Westerville, OH, 1986.
5. G.Y. Onoda, E.G. Liniger, and M.A. Janney, Dilatancy and Plasticity in Ceramic Particulate Bodies, pp611 in Ceramic Powder Science II, B. Edited by G.L. Messing, E.R. Fuller, Jr, and H. Hausner, American Ceramic Society, Westerville, OH, 1988.
6. W.H. Bauer and E.A. Collins, Thixotropy and Dilatancy, pp423 in Rheology Theory and Application, Vol 4, Edited by Eirich, Academic Press, New York, 1967.

CORRELATION OF CONFORMATION OF ACRYLIC POLYMERS IN AQUEOUS SUSPENSIONS AND PROPERTIES OF ALUMINA GREEN SHEETS

K. Nagata
Central Research Laboratory, Kyocera Corp., Kokubu. Kagoshima 899-43, Japan

ABSTRACT

The dispersion and rheological behavior of aqueous suspensions were investigated in relation to properties of alumina green sheets. A suspension made of polymer solution with pH 7.0 was well dispersed and showed Newtonian flow behavior due to the high density of adsorbed polymers. The strength and density of the green sheet were high in this case. In contrast a suspension made of polymer solution with pH 10.5 showed highly shear-thinning flow behavior due to low particle surface coverage, indicating the occurrence of flocculation. Both the strength and density of green sheet were lower in this case, while the strain of the green sheet was larger than that of the sheet prepared by the polymer solution with pH 7.0. The variation of polymer conformation at the alumina/water interface was discussed on the basis of above observations, the polymer conformation in water and the point of zero charge of alumina.

INTRODUCTION

Nonaqueous acrylic polymers are important binder additives and are widely used in industrial applications to prepare ceramic green sheets.[1-4] In contrast, aqueous acrylic polymers have been used minimally in tape casting. In this paper, the adsorption characteristics of these polymers on alumina are investigated and related to rheological behavior of suspensions and the properties of green sheets based on them.

EXPERIMENTAL

Materials

The polymer (P-1, LION Corp.) dissolved in solution is co-poly (acrylic ester and acrylic acid) resins with acid value of 35 mgKOH/g, Tg of -17 °C and pH of 7.0.

The polymer solutions of pH 7.5, 8.5, 9.5 and 10.5 are prepared by adding NH_3 solution to P-1. Solid contents of polymers in solutions are all adjusted to 25 wt%. Alumina is AL-45-1 from SHOWA DENKO Corp. The average diameter is 1.7 μm and the average surface area 2.0 m^2/g. The point of zero charge of this sample is 8.5.

EXPERIMENTAL METHODS

Preparation of suspensions and green sheets: 500 g of alumina, 100 g of water and 120 g of polymer solution are mixed for 20 h by a ball milling process. The pH of the suspension becomes 7.5 when polymer solution of pH 7.0 is mixed with alumina. Similarly suspensions of pH 8.1, 8.6, 9.4 and 10.4 are obtained by mixing polymer solutions of pH 7.5, 8.5, 9.5 and 10.5 with alumina respectively. The suspensions are cast on PET films and dried at 80° C for 1 h.

Strength and strain of green sheets: Strength and strain of green sheets are determined by means of an Instron-type machine.

Viscosity measurement: Viscosity measurement is conducted using a rotary viscometer model E, TOKYO KEIKI Corp. The ratio of viscosities at $5s^{-1}$ and $50s^{-1}$ of the suspension is taken as an index of flow behavior. Ratio of nearly unity means Newtonian flow behavior and relatively high ratio means shear-thinning flow behavior.

Measurements of molecular sizes of polymers: Molecular sizes of polymers of pH 7.0 and 10.5 are measured in water by use of the light scattering method, ASTEK Corp.

Measurement of adsorption: After the suspension is separated by centrifuge, the adsorbance is determined by measuring the concentration difference in the supernatant solutions.

RESULTS AND DISCUSSION

Packing Density of Green Sheet and Ratio of Viscosities at $5s^{-1}$ and $50s^{-1}$ versus pH of Suspensions

Figure 1 shows an almost linear decrease in packing densities of green sheets against the ratio of $5s^{-1}$ and $50s^{-1}$ of the suspensions. The suspension of pH 7.5 shows Newtonian flow behavior. In contrast the suspension of pH 10.4 shows highly shear-thinning flow behavior, indicating that the bridges between alumina and polymers at pH 10.4 cause flocculation.

Strength and Strain of Green Sheets versus pH of Suspension

Strength-strain curves of the green sheets are shown in Figure 2 as a function of pH

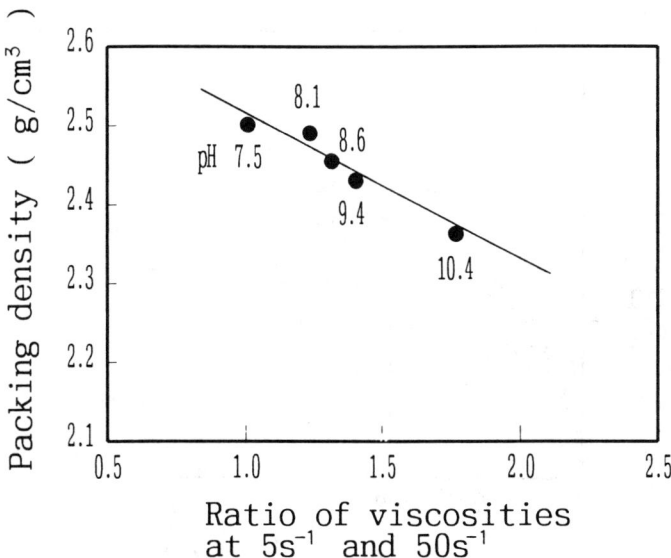

Figure 1 Packing density of green sheet and ratio of viscosities at 5s⁻¹ and 50s⁻¹ as a function of pH of suspension.

of suspension. As the pH of suspension decreases, the strength and packing density of green sheet increase, indicating that the strength is determined by the packing density of green sheet. With increase of pH of suspension the strain of green sheet increases. The green sheet obtained from suspension of pH 10.4 shows a plastic deformation under stress.

These phenomena can be explained by two types of interactions (1) between polymer and water, (2) between alumina and polymer.

<u>Molecular sizes of polymers in water</u>

Molecular sizes of polymers of pH 7.0 and pH 10.5 determined by the light scattering method are 3.6 nm and 50.1 nm. As the pH of polymer solution increases, the molecular size increases. It is clear that the polymer of pH 7.0 consists of dense, small coils because of the small number of negative surface sites. The polymer of pH 10.5 is expected to be in the form of large, expanded coils because of the electrostatic repulsion between negatively charged surface sites.[5-8]

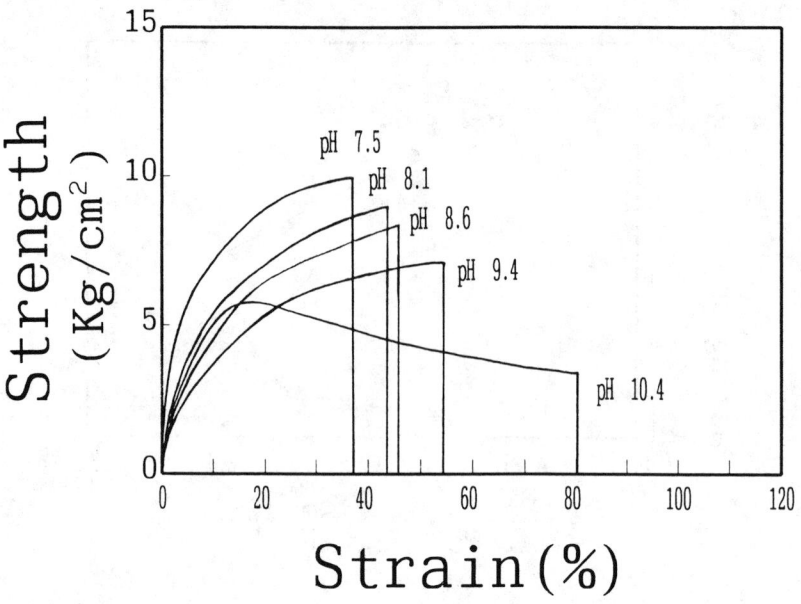

Figure 2 Strength-strain curve of green sheets.

Adsorption of polymer onto alumina versus pH of suspension

Figure 3 shows the adsorption of polymers on alumina as a function of pH of suspension. As the pH of suspension increases, the adsorption of polymers on alumina decreases. At pH < 8.5 the polymers are so small and so strongly adsorbed on the surface of alumina because of the electrostatic attraction between the positive charge on alumina and the negative charge on the polymer that the amount of adsorption is presumably large. At pH > 8.5 the surface of alumina has negative charge the same as the polymer. Then the electrostatic repulsion causes the polymer, which is in the form of relatively large, expanded coils, to be less adsorbed on the surface of alumina. The strain of green sheet increases with increasing pH of suspension. This behavior may be closely related to the large amount and the expanded form of free polymers with increasing pH of suspensions. Because free polymers in the solutions are deposited between alumina particles after drying in the tape casting process, the free polymers are considered to play like arms more than fixed polymers do. Taking the experimental results into consideration, the variation of polymer conformation in suspension and green sheet must be schematically shown in Figure 4.

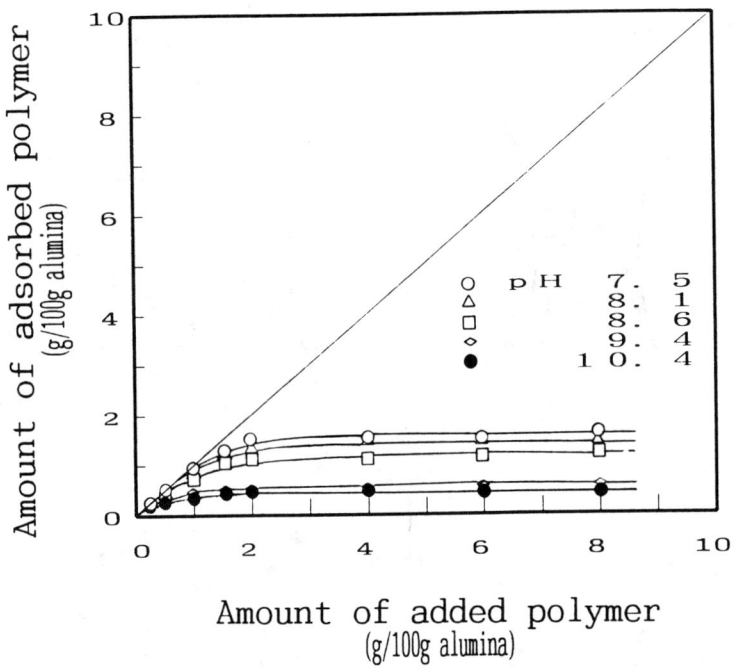

Figure 3 Adsorption of polymer on alumina.

CONCLUSIONS

Polymers adsorbed on the surface of alumina in the suspension of pH 7.5 are understood to be in flat and dense conformation because of the electrostatic attraction between polymer and alumina. In this case good dispersion is indicated by the Newtonian flow behavior and high packing density of green sheet. In contrast polymers are understood to be in the form hanging from alumina particles into the solution in the suspension of pH 10.4. Because of the electrostatic repulsion between both negatively charged polymers and alumina, the polymers are stretched chains and bridge alumina particles resulting in flocculation. Poor dispersions are indicated by the shear-thinning flow behavior and low packing density of the green sheet. As the volume of free polymers in the suspension increases, the strain of green sheet increases. Thus the amount and conformation of free polymers in the suspension are considered to play an important role in the strain of green sheet.

REFERENCES

1. M. D. Sacks and C. S. Khadikar, J. Am. Ceram. Soc., 66 [7] 488-494 (1983).
2. D. J. Shanefield and R. E. Mistler, Am. Ceram. Soc. Bull., 53 [5] 416-420 (1974).

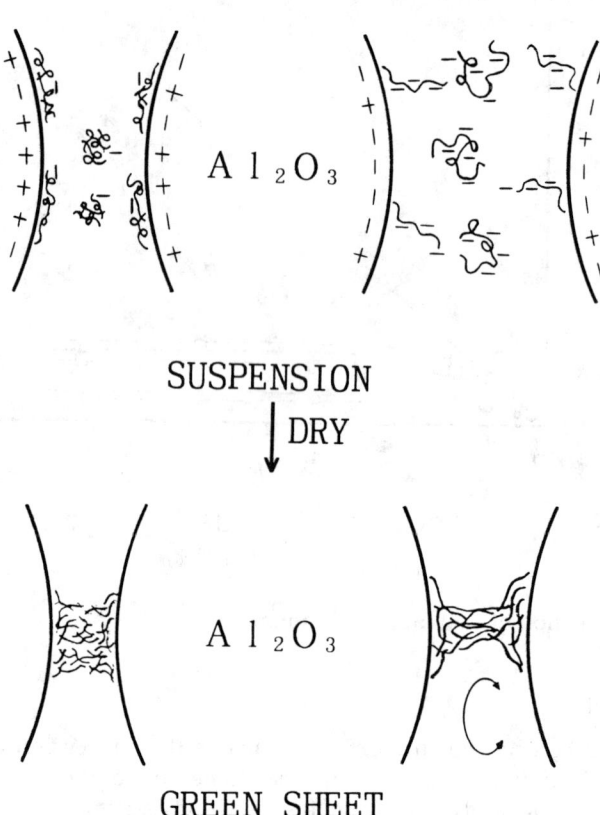

SUSPENSION

DRY

Al_2O_3

GREEN SHEET

Figure 4 Variation of polymer conformation in suspension and green sheet.

3. J. B. Blum and W. R. Cannon, Mat. Res. Soc. Symp. Proc., 40, 76-82 (1985).
4. L. B. Braun, J. R. Morris and W. R. Cannon, J. Am. Ceram. Soc., 64 [5] 727-729 (1985).
5. J. Cesarano. III, I. A. Aksay and A. Bleier, J. Am. Ceram. Soc., 71 [41] 250-255 (1989).
6. J. C. III and I. A. Aksay, J. Am. Ceram. Soc., 71 [12] 1062-1067 (1988).
7. K. Tjipangandjara, Y. Huang, P. Somasundaran and N. J. Turro, Colloid and Surface, 44, 229-236 (1990).
8. J. Papenhuijzen, H. A. Vander Schee and G. L. Fleer, J. Colloid Interface Sci., 104, 540-561 (1985).

HIGH PERFORMANCE ELECTRONIC SUBSTRATES FROM TAPE CASTING OF BERYLLIUM OXIDE

J. L. Sepulveda, and R. E. Kottman
Brush Wellman, Tucson, AZ 85706

ABSTRACT

Modern production of thin and thick film metallized beryllia devices requires thin, large area substrates with a highly smooth "as fired" surface. The production of substrates as large as 15x15 cm with a maximum thickness of 0.10 cm and with a surface finish of less than 0.13 μm center line average is accomplished by tape casting. High purity beryllia powder is well dispersed and finely ground in an organic medium using tumbling ball and Vibro-Energy mills prior to casting using a doctor blade. Dried green sheets of tape are then blanked, punched, and fired. Scoring, laser, ultrasonic or diamond scribing can be done on the fired ceramics. Most important operating variables to control the process and other economic considerations are discussed in detail.

INTRODUCTION

Tape casting of beryllia has been used for the production of flat substrates of up to 15x15 cm with thicknesses in the 0.013-0.102 cm range and camber of less than 0.003 cm/cm during the last ten years. This process exhibits technical and economical advantages for the production of large size substrates as compared to dry pressing.

Characteristics of beryllia tape are high thermal conductivity (TC = 290 W/mK) and smooth finish on an "as fired" surface (less than 0.13 μm CLA). These properties make beryllia tape substrates specially suited for the manufacturing of thin film metallized components, direct bond copper substrates, cofired packages, and post-fired packages.

Tape casting of beryllia involves slip preparation, slip conditioning, tape casting, drying, blanking, punching, and firing. Details of some of these processes are provided in this paper.

SLIP PREPARATION

A beryllia slip for tape casting consists of an organic solvent and beryllia powder containing a sintering aid, a binder, a plasticizer, and dispersants. The tape is cast on a silicone coated Mylar® sheet which is used as support.

Different slip formulations have been used for different applications. All of them contained polyvinyl butyral as binder and butyl benzyl phthalate or dibutyl phthalate as plasticizer. The solvent used consisted of a mixture of toluene/isopropanol, or MIBK/ethanol, or heptane/toluene, as the main components. Glycerol trioleate or Tergitol® NP-10 were used as dispersants.

The beryllia slip was prepared by using a series of deagglomeration steps which included shearing, ball milling, and Vibro-Energy milling. Additives were incorporated into the slip at different points of the process to control slip rheology. After deagglomeration the slip was passed through a screen and transferred to an agitated deairing tank where it was kept up until casting.

The smoothness of the surface after firing was directly related to the fineness of the beryllia used, which in turn was associated with the degree of deagglomeration. Similar behavior has been observed for the tape casting of alumina.[1-3] Beryllia powder being produced by chemical precipitation, consists of fine particulate material with crystallite size around 0.2-0.3 μm. Agglomeration occurs during calcination of the powder in the production cycle. The extent of the deagglomeration during the initial stages of slip preparation is well represented by the example shown in Figure 1. Median particle size in between 0.5 and 0.7 μm is typically used. This size distribution is similar to the one used in other beryllia manufacturing processes based on the pressing of spray dried powder.[4] Although this particle size produces a very smooth fired surface (typically between 0.076 and 0.102 μm CLA) and allows for good powder packing in the green body, it slows down the drying kinetics.

Characterization of slip rheology revealed a non-newtonian behavior characteristic of this type of fine particulate slip system. Figure 2 shows the shear thinning behavior of the slip when submitted to different degrees of agitation. Shear thinning and hysteresis shown in Figure 2 indicate that care has to be exercised in handling the slip during casting to attain consistent operation.

TAPE CASTING

The slip was cast on an NGK caster fitted with a 40.6 cm wide doctor blade. Casting speed was 30.5 cm/min. The capacity of this caster is estimated to be around 6,500 m^2/yr.

Green tape is slitted, blanked, and punched to generate parts of different geometries

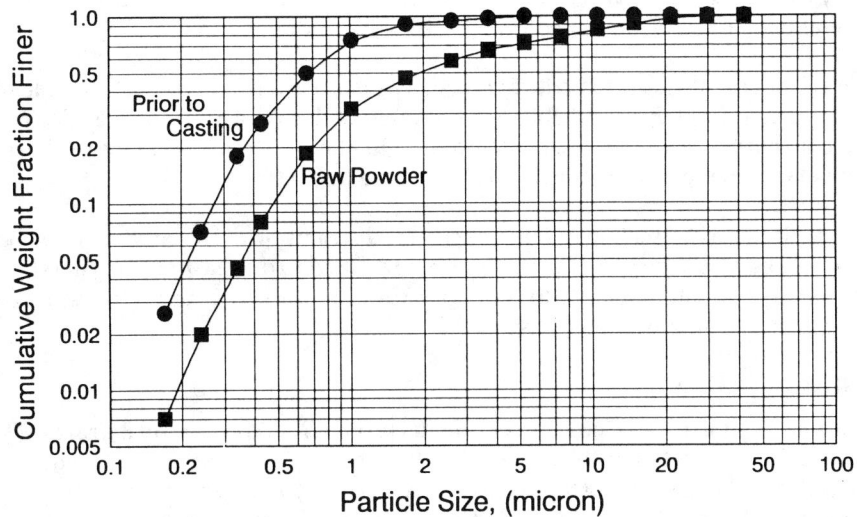

Figure 1 Beryllia particle size distribution prior to casting compared to raw powder size distribution.

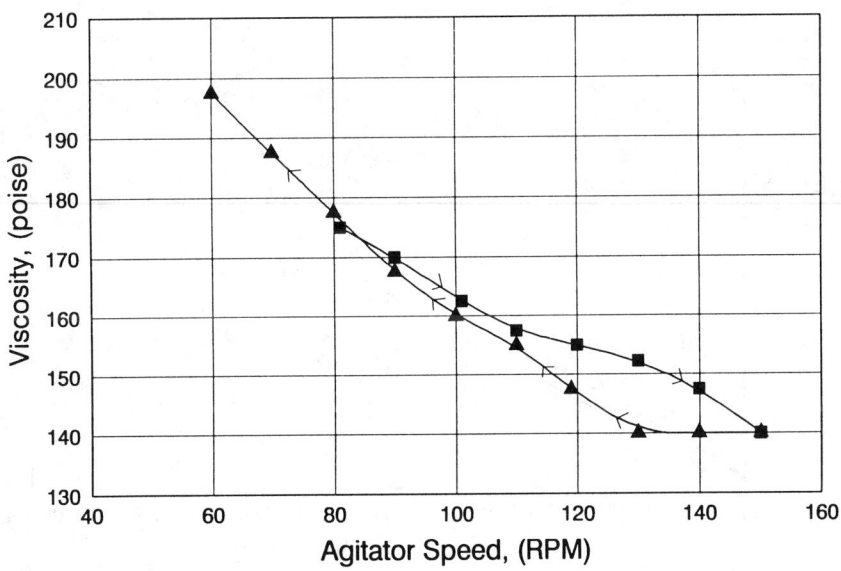

Figure 2 Slip viscosity versus agitator speed.

ready to be fired. Alternatively, several layers of green tape can be imprinted with a metallizing ink and laminated prior to firing to produce cofired packages.

Control of tape thickness is critical during casting. Figure 3 shows an example of the green and fired thickness obtained for different doctor blade positions. The fact that both curves can be represented by a straight line is indicative that the thickness shrink factor remains the same for different tape thicknesses.

Drying time was a function of the particle size distribution of the beryllia used and the thickness of the tape. Longer times were required to dry the finer beryllia slips. Thicker tape also took longer times to dry. Control of the drying cycle was a key parameter to insure product consistency.

FIRING

Tape was fired using periodic or continuous furnaces . A conventional firing cycle that reaches 1500°C in air was used. The furnaces used silicon carbide elements. Capacity available for firing tape is close to 970 m^2/yr. However, this capacity could be expanded easily to greater than 6,500 m^2/yr by allocating more kilns available in the plant to this operation.

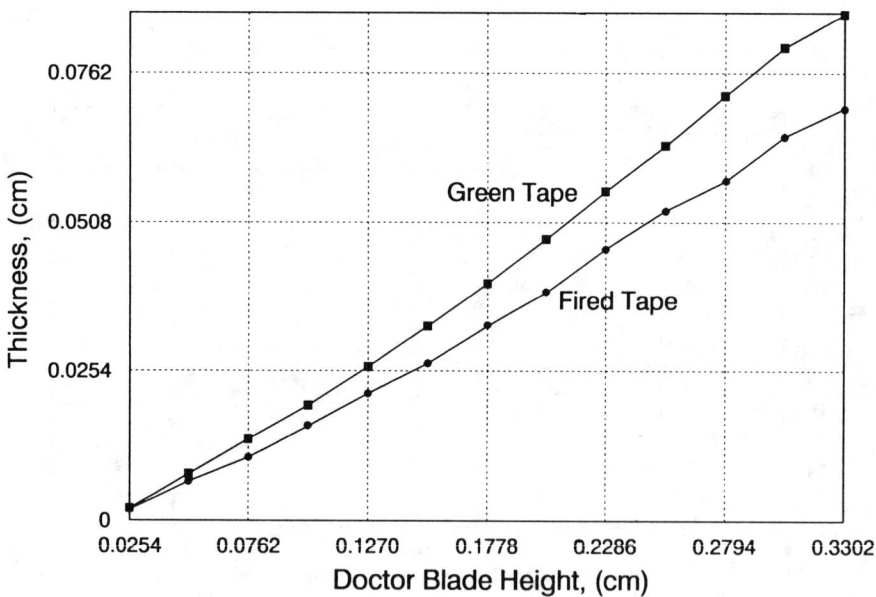

Figure 3 Green and fired tape thickness versus doctor blade height.

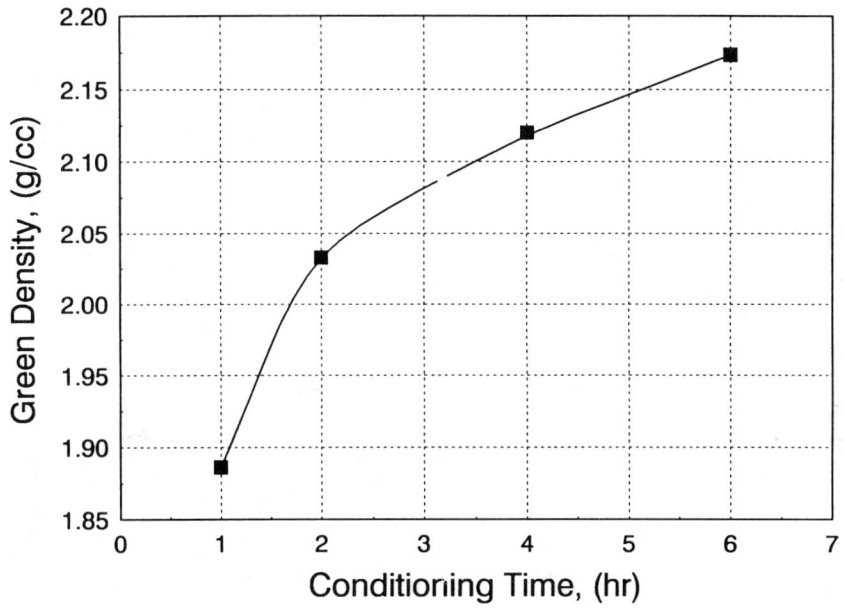

Figure 4 Effect of dispersant conditioning time on tape green density.

Figure 4 shows the effect of the dispersant conditioning time on green density. An increase in green density is observed for longer conditioning times. As expected, a decrease in shrinkage is obtained for longer conditioning times as shown in Figure 5. Fired density of the parts is kept at greater than 2.86 g/cc by controlling the firing cycle.

The main objective of a successful beryllia tape casting operation is to produce parts from lot to lot under controlled and predictable conditions to reduce variability. Statistical process control techniques are being used to achieve these objectives.[5] The improved yield derived from the application of these techniques directly impacts the economics of the process. Key parameters that affect quality and yields are: powder deagglomeration, slip rheology, green thickness control during casting, drying kinetics, and shrinkage and camber control during firing.

Experimental evidence collected during this study indicated that the surface finish of fired tape parts was directly related to the beryllia particle size distribution prior to casting and the degree of deagglomeration. Shrinkage during firing was controlled by adjusting the dispersant conditioning time while fired density was adjusted during the firing cycle.

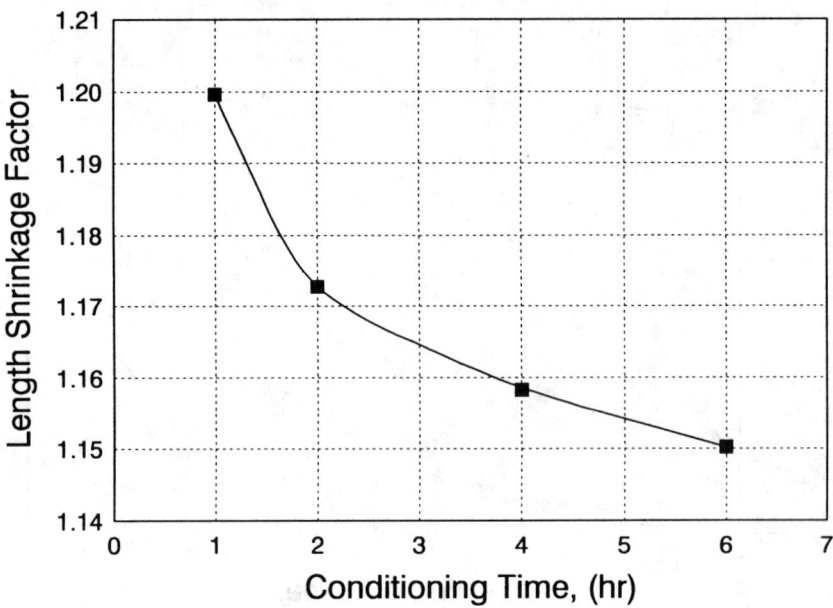

Figure 5 Effect of dispersant conditioning time on tape length shrinkage.

Product properties have been summarized in Table 1. Properties of tape substrates resemble those of other beryllia substrates. Tape surface finish, typically 0.076-0.102 µm CLA, falls between highly polished beryllia substrates which can be as low as 0.018-0.025 µm CLA and dry pressed substrates with 0.152-0.229 µm CLA finish.

Green scoring and punching coupled to laser scribing and drilling of beryllia tape substrates have been successfully performed on an industrial scale during the last years providing excellent process capabilities to produce large size arrays of electronic devices. Imprinting of arrays on large substrates is accomplished by thin film techniques or other metallization technology such as direct bond copper. Using beryllia tape substrates with an "as fired" surface of less than 0.13 µm CLA finish, a 20% increase in yield was obtained after imprinting arrays of devices by thin film metallization techniques as compared to dry pressed lapped or dry pressed lapped and polished substrates.

Diamond scribing of beryllia tape substrates has also been successfully demonstrated.[6] This technology provides a lower cost alternative to the dicing of arrays of parts after metallization.

Table 1 Tape Cast Substrate Properties.

PROPERTY	METHOD	TYPICAL VALUE
Thermal Conductivity	Laser Flash	290 W/mK
Thermal Coeff. Expan.	ASTM D-372	$6.4 \times 10^{-6}/°C$ 25-300°C $7.2 \times 10^{-6}/°C$ 25-500°C $8.1 \times 10^{-6}/°C$ 25-800°C $8.6 \times 10^{-6}/°C$ 25-1000°C
Dielectric Strength (0.025" thick)	ASTM D-116	770 V/mil
Dielectric Constant	ASTM D-150	6.5 - 1 KHz 6.5 - 1 GHz
Dissipation Factor (Loss Tangent)	ASTM D-150	1.1×10^{-4} - 1 KHz 1.0×10^{-4} - 1 GHz
Volume Resistivity	ASTM D-1829	4.5×10^{15} ohm-cm - 25°C
Density	ASTM D-373	> 2.86 g/cc
Surface Finish	Profilometer 0.076 cm Cutoff	< 0.13 μm CLA -Mylar® Face < 0.13 μm CLA -Air Side
Camber	Parallel Plates	< 0.003 cm/cm
Thickness	Micrometer	0.013-0.102 cm ± 10% NLT 0.0025 cm
Length, Width	Micrometer	15 cm max ± 10% NLT 0.0025 cm

CONCLUSIONS

Beryllia substrates of up to 15x15 cm, 0.013-0.102 cm thick, with camber of less than 0.003 cm/cm, and surface finish of less than 0.13 µm CLA were produced. Large area substrates with a smooth, "as fired" surface finish produced increased yields for thin film metallization applications. Also, direct bond copper was successfully applied to large fired tape substrates producing a strong and hermetic metal-ceramic bond.

217

Beryllia green tape imprinted with tungsten metallization ink was laminated and fired to produce hermetic multilayer cofired packages.

Surface finish of the fired tape parts was controlled by the powder particle size distribution and the degree of deagglomeration. A finer particle size distribution produced a smoother surface finish. Shrinkage was controlled by the dispersant conditioning time while fired density was controlled by the firing cycle.

REFERENCES

1. H. W. Stetson, and W. J. Gyurk, "Alumina Substrates", Western Electric Co., U.S. Patent 3,698,923., Oct. 17, 1972.
2. J. L. Park, "Manufacture of Ceramics", American Lava Corporation, U. S. Patent 2,966,719., Jan. 3, 1961.
3. D. J. Shanefield, and R. E. Mistler, "Fine Grained Alumina Substrates: I, the Manufacturing Process", Ceramic Bulletin, Vol. 53, No. 5, 1974.
4. J. L. Sepulveda, and D. A. Kahler, "Spray Drying of Beryllium Oxide Powder", 4-JXVI-91, 93rd Annual Meeting of The American Ceramic Society, Cincinnati, Ohio, April 29 - May 2, 1991.
5. G. P. Ferguson, D. E. Jech, and J. L. Sepulveda, "Statistical Process Control Applied to Manufacturing of Beryllia Ceramics", 71-SVIIP-91, 93rd Annual Meeting of The American Ceramic Society, Cincinnati, Ohio, April 29 - May 2, 1991.
6. T. L. Muir, and R. L. Flannigan, Villa Precision, Inc., Phoenix, Arizona, Nov. 27, 1990.

HIGH-SPEED PRECISE ELECTRON BEAM PERFORATION TECHNOLOGY

Kouichi Sakurai, Yoshio Yamane, Hidenobu Murakami,
Sigeo Sasaki and Shozui Takeno
Manufacturing Development Laboratory, Mitsubishi Electric Corporation, 8-1-1
Tsukaguchi-honmachi, Amagasaki Hyogo, Japan 661

ABSTRACT

We developed a high-speed and precise electron beam perforation system equipped with a rapid double-stage beam deflector and with a high-brightness EB gun. The system can make holes of 100μm in diameter through a 0.635-mm-thick sintered Al_2O_3 plate at a rate of 200 holes per minute.

INTRODUCTION

Demands for high speed and precise perforation have been increasing for ceramics as it has been acquiring wider usage in the electronic industry. The EB perforation technique has many advantages over the conventional methods, such as drilling, punching, and laser perforation especially for mass-production: They are (1) high machining rate (2) capability of small size perforation (3) processing after firing (4) no wear of machine tools.

Research on EB perforation has been reported for years. There has, however, been no EB system which can be used in practice because of low perforation speed and precision. We have designed EB system for both high quality and high perforation through-put. Development of a rapid double-stage beam deflector and a high-brightness EB gun has realized an EB perforation system which is applicable to practical uses.

APPLICATION OF EB TO PERFORATION

When a material is bombarded by EB, the kinetic energy of the beam is converted to the thermal energy causing the material to melt and evaporate. If beam irradiation is continued, electrons are scattered by the evaporated material which causes the beam and hole to broaden. This adverse condition is eliminated when the EB is pulsed with a duration time shorter than several hundred microseconds and a

219

repetition period longer than ten milliseconds. A substantial rise in vapor pressure does not take place until after several hundred microseconds after the beam irradiation and because the vapor diffuses away in a time of ten milliseconds. A hole is made through by repetition of several EB pulses. This process, which we call "multi-EB-pulse perforation", is required for high quality perforation[1] and is adopted in our system.

DEVELOPMENT OF EB PERFORATION SYSTEM

High Through-put Multi-pulse Perforation

In order to achieve a high through-put in multi-EB-pulse perforation processes, efficient use of beam off-time is necessary. The beam off-time is relatively long compared with actual irradiation time because the vapor must be allowed to diffuse away.

EB can be deflected rapidly by an electromagnetic force to a required position on the workpiece-surface. Thus, the beam can be moved to the location of another hole and its perforation is conducted by one pulse irradiation during the waiting time for the vapor diffusion of the preceding hole. Generally, more than ten holes can be pulse-irradiated during one diffusion time of the vapor. In our system this sequence is controlled by a computer as shown in Fig. 1. As a consequence, this "high through-put multi-EB-pulse perforation" attains the precise perforation together with a high through-put.

We developed a rapid double-stage beam deflector and a high brightness EB gun in order to realize "high through-put multi-EB-pulse perforation".

Rapid Double-Stage Deflector

If the EB is deflected with a single-stage deflector like in a conventional CRT, then the EB is not at right angles to the workpiece-surface and perforated hole is inclined. We developed a double-stage deflector which allows a vertical beam irradiation. Magnetic characteristics of a deflection coil column were improved by reducing the eddy current on the column. This improvement resulted in a high speed beam scanning of 2 km/second.[2]

We also developed a technique to reduce beam-deflection aberrations such as distortion or astigmatism. The computer for scanning data generation also corrects the distortion, so that the beam positioning error is reduced to ±20μm. The hole diameter accuracy is controlled to better than ±10% by a stigmator and by a stabilized electric power source for the beam current.

220

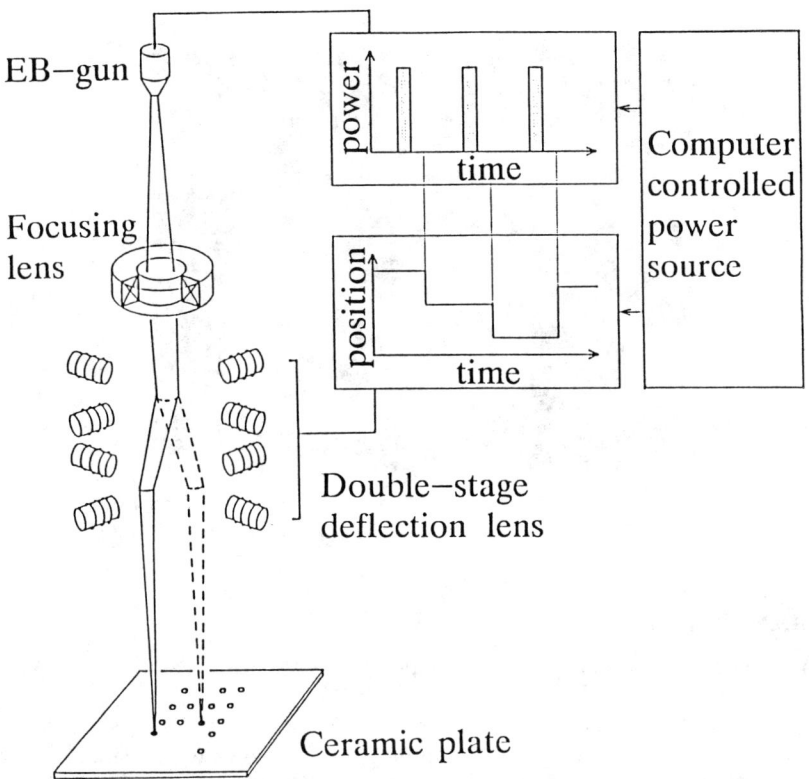

Figure 1 Method to achieve high-speed machining. A rapid double-stage beam deflector and a high-brightness EB gun are controlled by the computer.

High-Brightness EB Gun

A high-brightness EB gun and its optimized focusing system were developed to obtain a fine beam with a high intensity and sharp energy distribution. The components were designed with the help of computer simulation and succeeded in obtaining a beam with a high energy density ($10^7 W/cm^2$) at a low, in other words practical, accelerating voltage (60kV). The volume of material removed per one pulse-irradiation increases with the energy density of the beam, as the number of pulses required to make the perforation also decreases.

Thanks to these improvements, the maximum machining rate reached the order of a hundred holes per minute and seems to be only limited by the fracture toughness of ceramic against the thermal shock of EB irradiation.

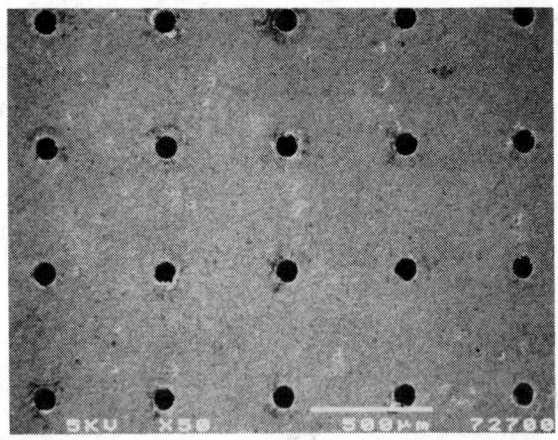

Figure 2 The outside view of the EB-perforated 100-μm-diameter holes for 0.635-mm-thick Al₂O₃ sintered plate. 200 holes can be processed in a minute.

PROCESSED EXAMPLES

Some typical examples of perforation experiments are described below.

Fig. 2 shows 100-μm-diameter holes for a 0.635-mm-thick Al₂O₃ sintered plate processed at a rate of 200 holes per minute. The outside view of a small (30μm in diameter, 0.1mm in depth) hole is shown in Fig. 3. This system can also make holes with a high aspect-ratio (25, 100 μm in diameter, 2.5mm in depth). Most oxides, such as Al₂O₃, tend to be melt and evaporate. Some portion of the molten material deposits on the internal surface of a hole thinly but it is easily removed by a post-process treatment.

Fig. 4 shows a cross section of a 100-μm-diameter hole through a 1.0-mm-thick Si₃N₄ plate. The machining rate is 200 holes per minute. Most kinds of carbides and nitrides such as SiC and Si₃N₄, are "sublimable" (decomposed or evaporated). For these sublimable ceramics, deposition on the internal surface was very little.

CONCLUSIONS

We developed an EB system which is equipped with a rapid double-stage beam deflector and with a high-brightness EB gun. The system can perforate precise holes (smaller than 100μm in diameter) at a high machining rate (more than 200 holes per minute) by what we call "high through-put multi-EB-pulse perforation" method. We consider this system and technique to be applicable to many materials and applications.

222

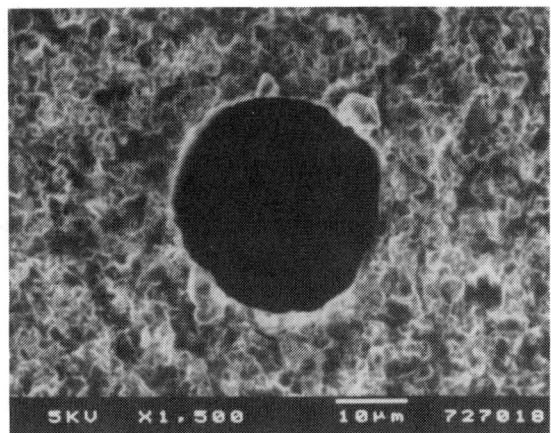

Figure 3 The outside view of the 30-μm-diameter hole for 0.1-mm-thick Al_2O_3 sintered plate.

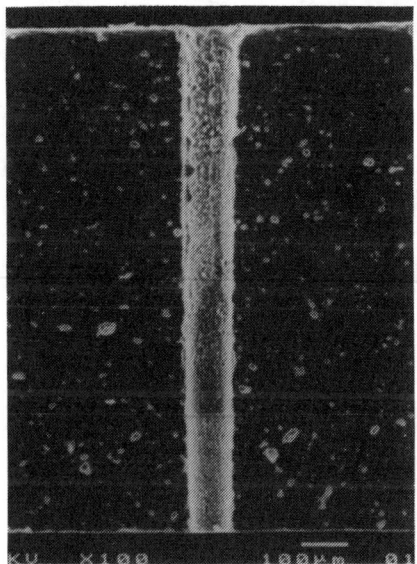

Figure 4 The cross section of a 100-μm-diameter hold through 1.0-mm-thick Si_3N_4 plate as an operative example of "sublimable" material. The machining rate is 200 holes per minute.

REFERENCES

1. S. Hiramoto, Y. Yamane, K. Sakurai, M. Ohmine "A study on energy beam drilling (2nd report)", Journal of High Temperature Soc., vol.17, No.1, pp.27-33 (1991).
2. T. Iwami, M. Sakamoto, H. Murakami, S. Sasaki, S. Hoshinouchi "Large area electron beam direct imaging technology for printed wiring boards", In Proc. of 15th Annual Conf. of IEEE Industrial Electronics Soc., vol.3, pp.550-555 (1989).

A METHOD FOR MAKING GROOVES WITH SHARP CORNERS ON A GREEN CERAMIC BODY USING A TOOL WITH BIAXIAL ULTRASONIC VIBRATION

Kiyoshi Suzuki and Hiroshi Nakabayashi
Nippon Institute of Technology, Saitama, Japan

Tetsutaro Uematsu
Toyama Prefectural University, Toyama, Japan

Shoji Mishiro
Taga Electric Co. Ltd., Tokyo, Japan

ABSTRACT

A new technique for making grooves or slits with sharp inside corners on a green ceramic body has been developed. Such grooves can not be machined with the existing machining technique. A specially designed vibrating device, which generates biaxial ultrasonic vibration of a blade-like metal bonded diamond tool, can successfully machine such grooves on a green Al_2O_3 block with rather high efficiency. Various types of grooves such as a straight groove and a 3-D shaped groove using NC program have been machined. No defect such as crack or distortion was observed on the ceramic body after sintering.

INTRODUCTION

Most of engineering ceramic parts are manufactured mainly by using molds with complicated cavities. Three typical methods for making ceramic parts with molds are a powder compacting method with a metal mold, an injection molding method with a metal mold, and a slip casting method with a porous plaster mold.

Some of these green parts require subsequent machining process before sintering. Machining of green ceramic material is usually done with cutting tools and the machining efficiency is rather high in making flat surface, round bars or round holes. But such tools are not suitable for making a groove with sharp inside corners, because the corners made with a rotational cutter are obviously concave.

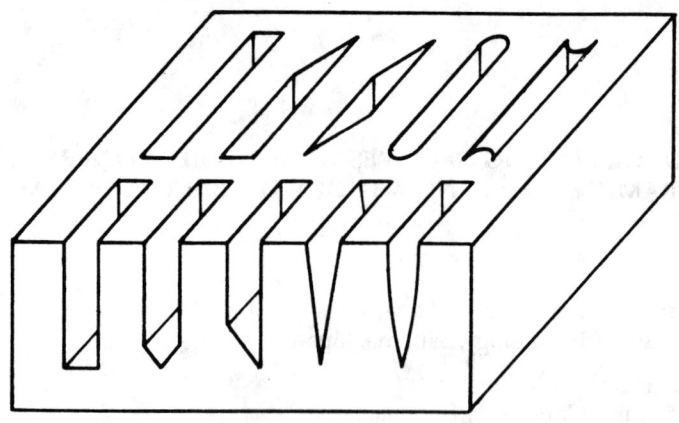

Figure 1 Shapes of grooves which cannot be machined with rotational cutting tools.

Even if such grooves as in Fig. 1, which do not permit corner radius or which require non-symmetrical depth with respect to the center line of the groove width, are desirable, there is currently no acceptable means of manufacture of these grooves.

In this paper, a new method for making grooves with sharp inside corners on a green ceramic body is proposed and its machining ability is investigated.

PROPOSAL OF A NEW METHOD FOR FORMING RIB HOLES

A new method for making a groove with sharp inside corners on a green ceramic body has been developed by the authors. In the method a blade-like metal bonded diamond tool is used instead of a rotational cutting tool and, as Fig. 2 shows, bending vibration and longitudinal vibration are given to the blade-like tool at the same time by a biaxial ultrasonic vibrating transducer.

Fig. 3 shows the construction of the developed forming device which can be mounted on a conventional NC machine tool such as an NC milling machine. The frequency and the amplitude of the tool at the longitudinal vibration mode are 28kHz and 20μm p-p, respectively. In case of the bending vibration mode, they are 19kHz and 12μm p-p, respectively.

FORMING EXPERIMENTS

Effects of Biaxial Ultrasonic Vibration

Experiments for making grooves (width 2mm) on green alumina ceramic bodies

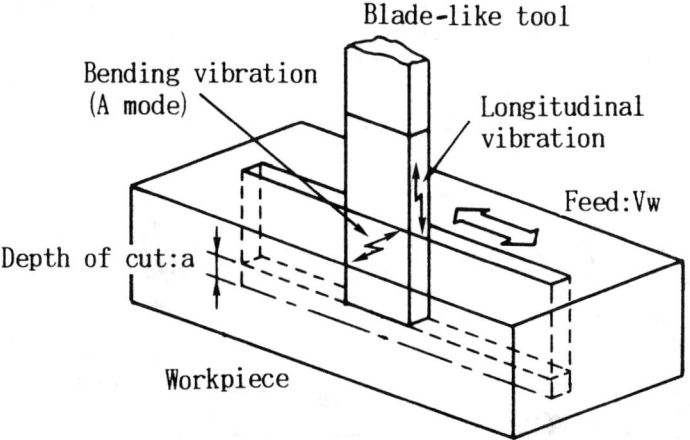

Figure 2 Method for forming a rib hole with biaxial ultrasonic vibration tool.

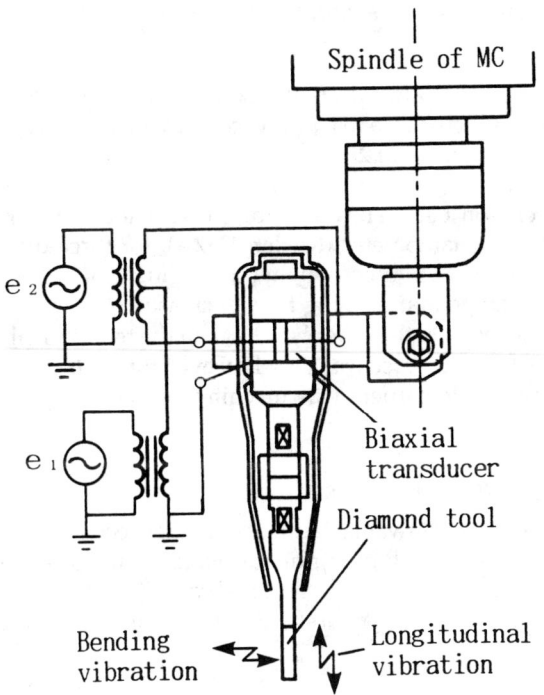

Figure 3 Schematic illustration of the developed biaxial ultrasonic device mounted on an NC machine tool.

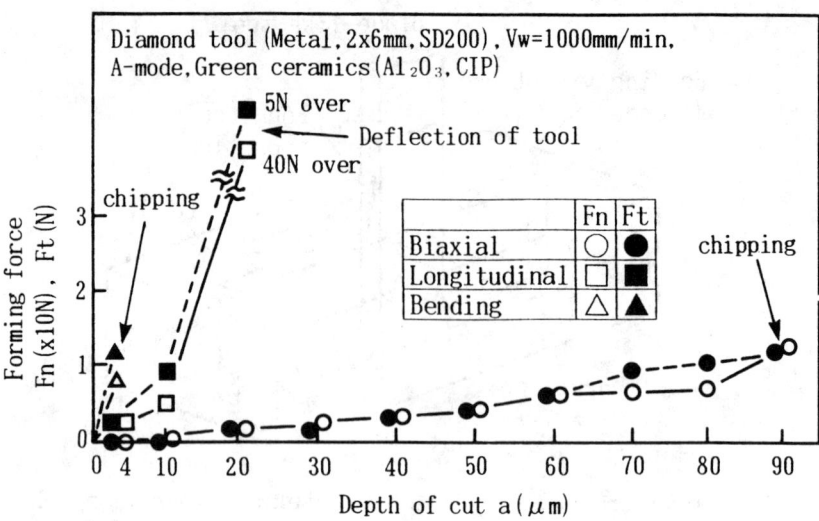

Figure 4 Forming force at various vibration modes.

(Sodick co. ltd. SA610, with 90% Al_2O_3 and some binders such as MgO, SiO_2 etc. made by CIP) were conducted on a machining center with a metal bonded diamond tool (SD200) with the thickness of 2mm.

Fig. 4 shows the relation between depth of cut and forming force, which was measured with a tool dynamometer (Kistler 9257A), for three different vibration modes. When the tool was vibrated only in the longitudinal or bending direction, both the normal and tangential forming force increased steeply with the depth of cut and loading occurred. On the other hand, when the tool was vibrated biaxially at the same time, the forming force was very low even for the large depth of cut and a groove with sharp inside corners was machined successfully with rather high stock removal rate.

Optimum Forming Conditions

Fig. 5 shows the maximum allowable depth of cut in respect to feed rate for making grooves successfully without chipping or breakage. The figures in the brackets indicate the forming efficiency or stock removal rate Z', and it will be understood that adoption of higher feed rate is preferable for achieving higher forming efficiency.

The maximum stock removal rate in the experiments was $Z = 500 \text{ mm}^3/\text{min}(Z' = 250 \text{ mm}^3/\text{mm} \cdot \text{min})$ under the conditions of the feed rate of 5000 mm/min and the depth of cut of 50 μm. The time required for making a groove with 2mm width,

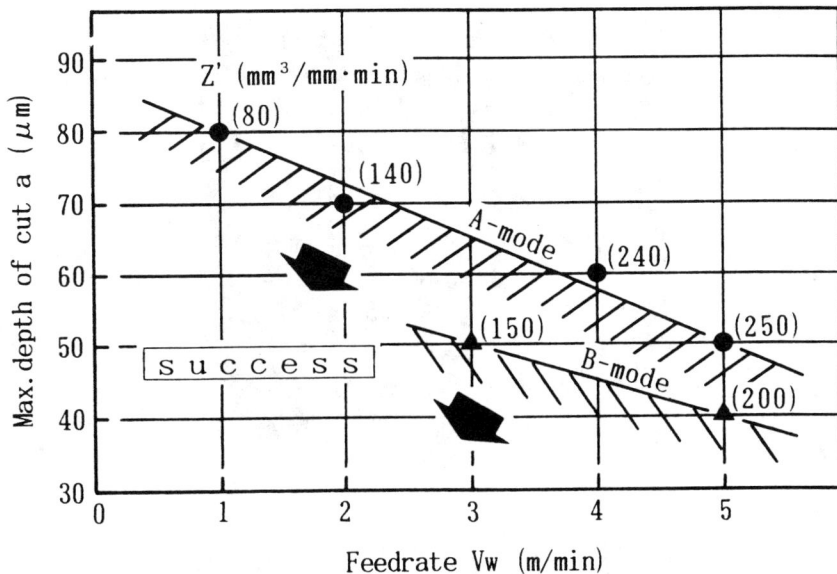

Figure 5 A region for successful forming without chipping or breakage.

36mm length and 5mm depth on a green alumina ceramic body was only 3 minutes.

VARIOUS SAMPLES FORMED BY THE METHOD

Fig. 6 shows various groove samples formed on green alumina ceramic bodies with the developed method and device. Fig. 6(a) is a sample with rib holes. Fig. 6(b) is a sample with stepped holes having sharp inside corners. A very narrow bridge with the width of 0.2mm was successfully formed between adjacent rib holes by selecting appropriate conditions.

CONCLUSIONS

To cope with forming of rib holes with sharp inside corners, which can not be formed or machined with existing rotational tools, a new forming method utilizing a biaxial ultrasonic vibration tool has been developed. Through various forming experiments on green alumina ceramic bodies, it has been found that the method is practically effective for such a shape.

ACKNOWLEDGMENTS

The authors would like to express their sincere thanks to Messrs. Yamazaki Mazak Co. Ltd., Sodick Co. Ltd. and Ikegami Mold Engineering Co. Ltd.

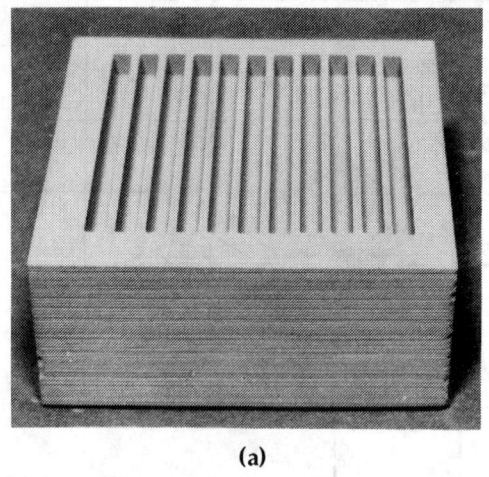

(a)

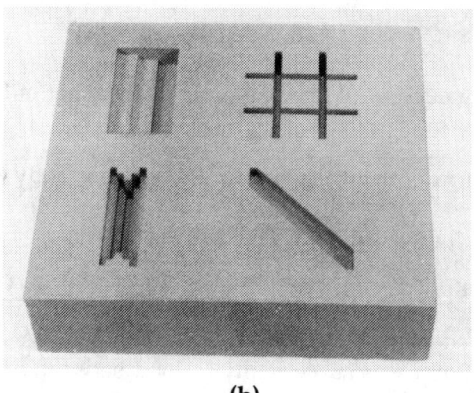

(b)

Figure 6 (a) Rib holes with sharp inside corners formed on a green Al_2O_3 body and (b) stepped holes with sharp inside corners (green Al_2O_3).

REFERENCES

K. Suzuki, T. Uematsu et al.: Attempt of forming rib holes on a green ceramic body with a biaxial ultrasonic vibration tool (in Japanese), Preprint of JSPE's Autumn Conf. 1990-10, 263.

OPTIMIZING GREEN MACHINING AFTER ISOPRESSING OF BERYLLIA CERAMIC BODIES

G. A. Kovell, and J. L. Sepulveda
Brush Wellman, Tucson, AZ

ABSTRACT

Modern production standards of beryllia ceramics requires machining the green body to near shape to optimize powder yields, reduce fired machining, and reduce cost. Practical experience gathered with standard machining as well as computer assisted CNC techniques are described. In addition, the mechanics of the isopressing of a beryllia green body, dimensional control, green machining, and dimensional control after firing are discussed. Application of the technology to the production of laser bores and other ceramic bodies is explained in detail.

INTRODUCTION

Cold isostatic pressing (CIP) has long been established as a proven method of consolidating powders into a given shape, particularly when the required shape has internal and/or external profiles. Brush Wellman has used the CIP technique to form beryllia parts from spray dried powders for many years. Several advantages derived from the use of this technique could be listed:

1) consistent green densities throughout the body yield consistent fired properties

2) large and/or complex shapes can be produced

3) internal and external features can be incorporated to the green body

4) near net shape parts can be formed minimizing fired machining costs

Equipment used to produce the parts discussed in this paper consisted of a National Forge Isomax 30 cold isostatic press with 4970 MPa(30 KPSI) capability with a 305 mm diameter x 914 mm long chamber. Typical part shapes generated in an industrial scale are:

1) blocks and discs	(178 mm maximum diagonal x 711 mm long)
2) tubes and crucibles	(.51 mm to 152 mm ID)
3) rods	(high aspect ratio)
4) laser bores	(multi hole tubes)
5) spark plugs	(profiled ID and OD)

In recent years, there has been a lot of discussion related to safety in handling beryllia. The handling of fired beryllia bodies is completely safe.[1] However, processes involving the handling of dry powder or the machining of green bodies which may generate airborne particulates require special attention.

During the forming process, all handling of beryllia is carried out in a safe manner by use of ventilation, with absolute filtration capability, in all operations where powder may become airborne. The handling of beryllia is safe as long as the concentration of airborne beryllia particles is less than the required OSHA standard of $2 \, \mu g/m^3$ of air.[2] Our standard operating concentration is maintained at less than $1 \, \mu g/m^3$.

FORMING

The general process flow is illustrated in Figure 1. The isopressing portion of it consists of:

1) loading the mold, usually made out of polyurethane or natural latex rubber, with powder and sealing it. De-airing is required on low green strength materials.

2) placing the mold into the isopress and sealing the vessel.

3) build up high pressure in the vessel

4) dwell at peak pressure

5) decompress

6) open isopress and remove mold

7) remove compact from mold

To minimize excessive powder usage and minimize green machining it is fundamental to have consistent mold filling. This is accomplished by using a computerized weigh feeding system that minimizes variability to less than ± 0.1 gram. At the same time, proper loading of the mold, by using vibration, will insure constant

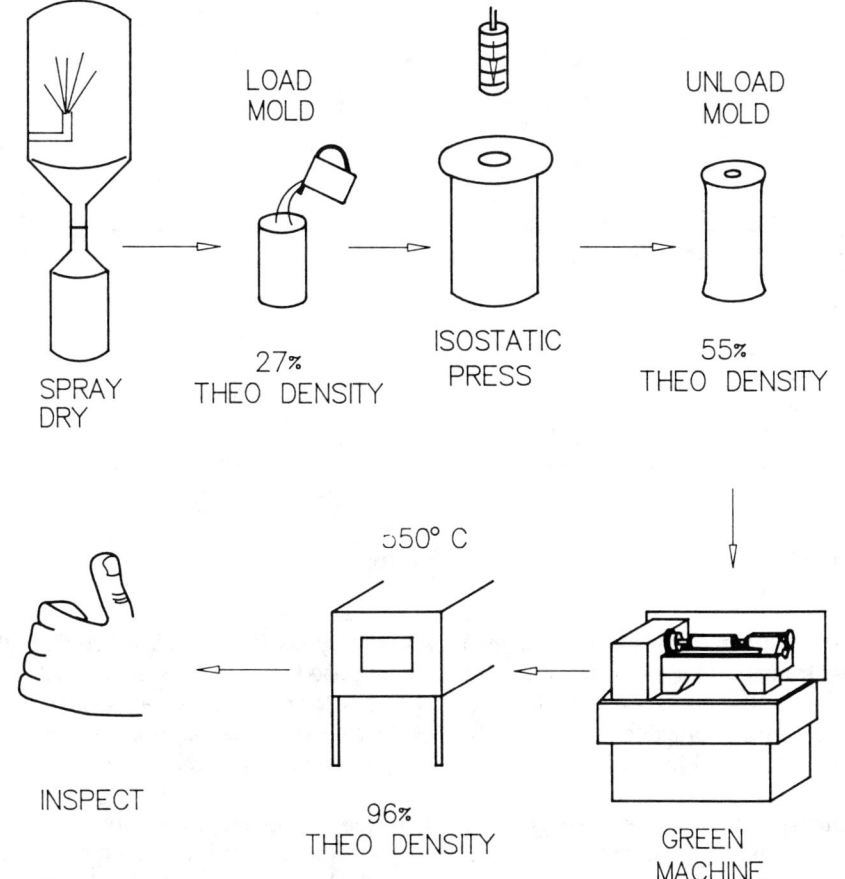

Figure 1 Isopress/green machine flow.

packing density of the powder which insures dimensional consistency of the compact.

One objective of isopressing is to achieve a green density that is uniform throughout the green body and suitably high to achieve a workable green strength as well as the required sintered density.

Figure 2 shows a plot of green density vs. dwell time at 3000 MPa peak pressure and room temperature. It indicates that dwell time has a pronounced effect on green density initially and a relatively smaller effect after extended dwell times. Figure 3, shows that temperature of the powder, at 3000 MPa and 20 sec dwell, also has an

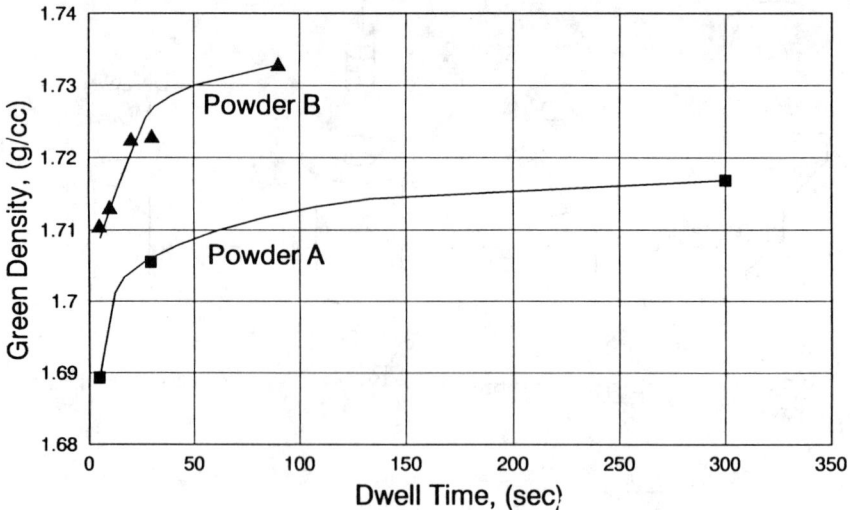

Figure 2 Effect of dwell time on green density.

effect on green density. This has been discussed by DiMilia[3] and is an effect of the glass transition temperature of the binder being used. For the system described in this paper, the necessity to control temperature at forming, as well as dwell time, is easily seen. In practice, this is accomplished by keeping the loading station under temperature and humidity control and using an automatic dwell timer.

Dimensional tolerances are typically ± 6% for the as pressed parts. However, ± 2% can be achieved through careful design and use of tooling, and controlling compression ratio of the powder. Surface finish characteristics of the parts will be affected by the surface finish of the tooling used, the spray dry powder size distribution, the degree of deagglomeration of the slurry prior to spray drying, and the raw beryllia particle size distribution.

GREEN MACHINING

When the as pressed tolerances are not acceptable to the final part requirements, green machining is used to allow tolerances ± 1% (1/2% typical). Cost effective advantages of green machining can be summarized as follows :

1) faster stock removal compared to post fired machining

2) allows near net shapes as fired (±1 %) minimizing or eliminating post fired machining

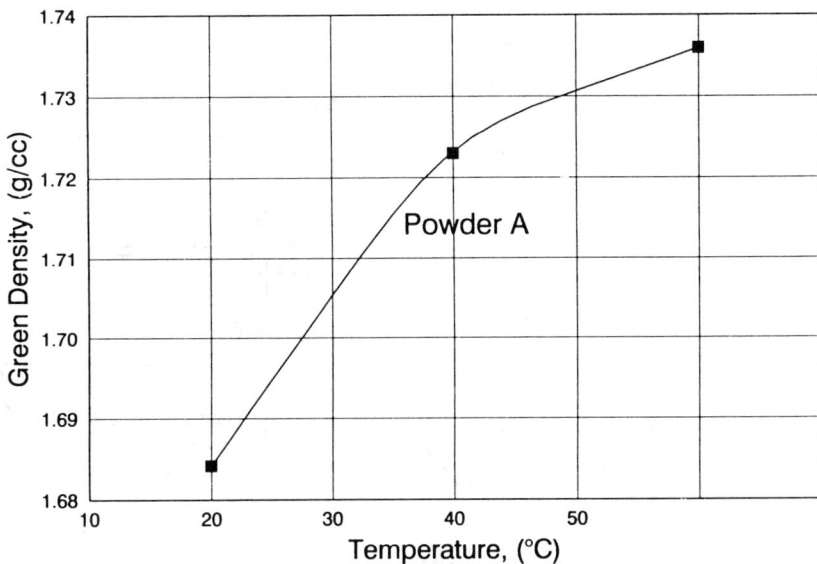

Figure 3 Effect of temperature on green density.

3) allows near net shape even in small lot sizes. Alternatively, high isopressing tool cost can make isopressing to near net shape expensive

Examples of the variety of features that can be green machined are shown in Figure 4. Most any lathe or mill operation can be performed in the green state as long as a few requirements are maintained:

1) hold the workpiece in the machine tool without damaging it. Bending, crushing, and cracking are all possible problems. This requires:

   a) relatively high green strength of the workpiece
   b) using minimum chucking pressure (1-2 MPa max)
   c) use vacuum arbors to mount workpiece where possible
   d) use rubber lined jaws or fixtures

2) make the beryllia particles machined away more capturable by minimizing their mass. Thus, ventilation systems with 250 to 450 meters per minute capture velocities, can capture airborne beryllia from the cuttings. This can be accomplished by establishing a balance between the material removal rate and the ability of the ventilation system to capture the cuttings. This must also be tempered with the need to get the job done as quickly as

235

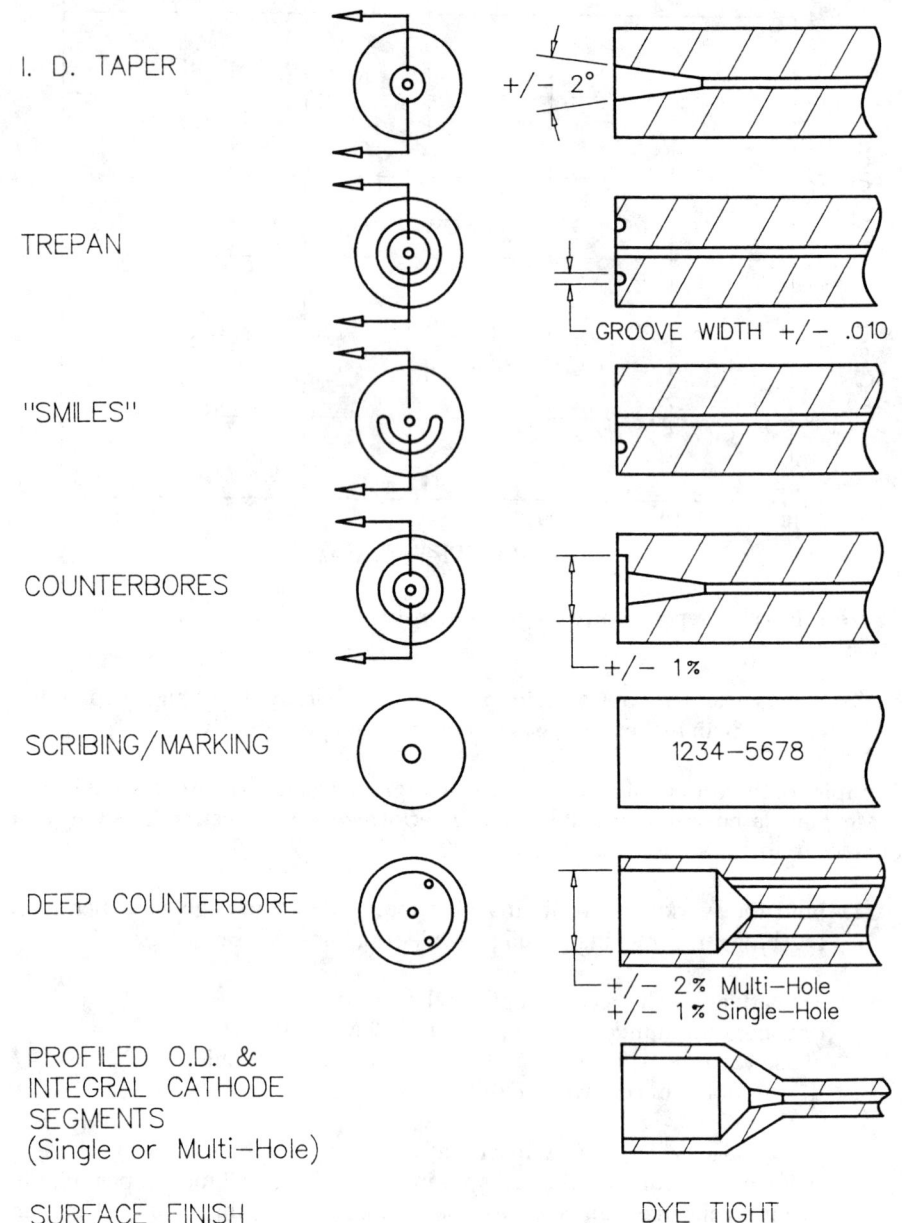

I. D. TAPER

+/- 2°

TREPAN

GROOVE WIDTH +/- .010

"SMILES"

COUNTERBORES

+/- 1%

SCRIBING/MARKING

1234-5678

DEEP COUNTERBORE

+/- 2% Multi-Hole
+/- 1% Single-Hole

PROFILED O.D. &
INTEGRAL CATHODE
SEGMENTS
(Single or Multi-Hole)

SURFACE FINISH   DYE TIGHT

Figure 4 As green machined shapes.

possible to minimize operating cost. Key parameters to control are:

a) minimize cutting speed (surface meters per minute)
b) minimize cutting tool infeed rate (cm per revolution)
c) minimize depth of cut

3) use diamond tipped cutting tools to resist the abrasiveness of the ceramic. Brush Wellman has found that even tungsten carbide tools do not remain sharp for more than a few parts.

In most cases, run sizes are moderate in size but even for these cases, the machine tool should be selected to minimize cost. A choice is made between conventional or computer assisted machining (CNC). The machine tools most commonly used for green machining of beryllia at Brush Wellman are:

1) Standard engine lathes

2) CNC lathe

3) Standard end mills

4) CNC end mill

Brush Wellman has found the following advantages of using standard machine tools to green machine :

a) best for simple operations

b) equipment is relatively inexpensive

c) best for small runs where simple and fast set ups are all that is required

Using CNC equipment, other advantages can be gained:

a) very repetitive (high process capability index[4], C_{pk} = 6.4 in a spark plug machining operation)

b) very tight tolerances can be easily held. Typically 7 microns repeatability is maintained.

c) higher production rates can be obtained. For example, one milling operation went from 10 minutes in a standard mill, down to 30 seconds in a CNC mill. This is especially true when the job requires a multi tool set up.

d) best when complex profiles are required on either ID or OD.

FIRING CONTROL

Once the parts have been green machined, they are ready to fire. At this time several factors must be considered:

1) Shrinkage Variation. This is controlled by the compact green density, which is driven by the pressure used to form the part.

2) Distortion and Warpage. Enough machining stock must be left so that the part will clean up during post fired machining.

3) Grain Size. Typically, this is controlled by choosing the correct kiln temperature profile and by controlling variations caused by the placement of the parts in the kiln.

4) Binder Bakeout. During the initial stages of the cycle, binder is completely removed. Parts may crack if the debinding rate is excessive.

The primary method of controlling the firing is through the use of control discs. These are dry pressed discs of known green density (within ± 0.1 %) and shrinkage for a given kiln cycle. These are placed throughout the kiln. Statistical Process Control techniques are used to monitor shrinkage and grain size of the discs to control the process.[5] This has the effect of determining if the kiln itself was functioning properly during the cycle or if special cause variation was present.

The secondary means of controlling the firing is to use curves of forming pressure versus shrinkage for a given powder lot. This allows one to set the forming pressure to match the design shrinkage of the tooling used. However, this effort can be minimized when the variation in shrinkage between powder lots is minimal. At Brush Wellman, one pressure is assigned to a given tool and rarely requires changing.

CONCLUSIONS

Isopressing coupled with green machining yields near net shape products. Economical advantages can be gained using conventional machine tools and further enhanced by applying CNC equipment. This technique has been successfully applied to the manufacturing of beryllia ceramic on an industrial scale for several years, resulting in cost reduction.

Key parameters to control the process are green density and strength, workpiece handling, appropriate dust control, and controlled firing.

It was determined that the effects of dwell time and forming temperature on green density were significant and must be addressed.

Throughout the manufacturing of beryllia parts, a safe working environment has been maintained. This was achieved by using high velocity localized ventilation.

REFERENCES

1. Holler III, James W., Colonel, USAF, Director, Engineering, "Beryllium Oxide (BeO) in DoD Managed or Procured Electrical or Electronic Items", Defense Logistics Agency, Defense Electronics Supply Center, Dayton, Ohio, March 12, 1991.
2. 29 Code of Federal Regulations, Part 1910.1000, Table Z-2.
3. DiMilia, R. A., and Reed, J. S., "Dependence of Compaction on the Glass Transition Temperature of the Binder Phase", Amer. Cer. Soc. Bull, 62[n4] 484-8, 1983.
4. Kane, V. E., "Process Capability Indices", Journal of Quality Technology, Vol. 18, Jan. 1986, pp. 41-52.
5. G. P. Ferguson, D. E. Jech, and J. L. Sepulveda, "Statistical Process Control Applied to Manufacturing of Beryllia Ceramics", 93rd Annual Meeting of The American Ceramic Society, Cincinnati, Ohio, April 29 - May 2, 1991.

RELIABLE ELECTROPHORETIC MOBILITY MEASUREMENT FOR CERAMIC POWDERS

Jiun-Fang Wang, Richard E. Riman, and Daniel J. Shanefield
Department of Ceramics, Rutgers University, Piscataway, NJ 08855-0909

ABSTRACT

Microelectrophoresis (ME) is an important technique for measuring surface properties of colloids. In order to obtain reliable measurements, a good reference colloid must be chosen first, and both the behavior of the reference under a variety of measurement conditions and instrumental factors must be established. Polystyrene (PS) latex has proven to be a good reference material. Time-dependence, solids-loading, particle size, and cell profile effects were found to affect the reproducibility. Using optimized referencing conditions, surface properties of silicon nitride were studied as a function of both aging conditions and surface modifications using aluminum and yttrium compounds.

INTRODUCTION

Ceramics having a reduced number of defects can be fabricated by using a suspension that is free of agglomerates. This can be accomplished by the use of electrostatically and/or sterically stabilized suspensions in forming processes such as slip casting. Agglomerate-free, stabilized suspensions are more readily obtained when the surface potential, ψ, is maximized (>25 mV). The surface potential can be inferred from calculation of the zeta potential, ζ, a measurement proportional to the electrophoretic mobility, μ. Thus, measurement of μ (defined as the drift velocity divided by the applied electric field) via commercially available microelectrophoresis (ME) instrumentation can be used to indicate the optimum conditions for suspension preparation.[1] However, for many ceramic materials, the isoelectric point (IEP), the pH at which μ is zero, is reported instead.[2]

A common problem encountered in consulting the ceramic colloids literature is the wide range of variability of IEP data for a given material.[3] This variation can be attributed to powder preparation techniques, sample history, suspension preparation methods, and/or equipment used to make the measurement. For silicon nitride powders, for instance, the IEP can range from 3 to 9 depending on the manufacturer, preparation history, and the conditions under which it is stored.[4,5] In addition, the relative contribution of the different ME instruments to variations in IEP data has not been assessed.

Many of the above problems could be solved if a universal reference colloid were selected for calibrating instruments prior to measurement of sample μ and a standard procedure were developed for sample preparation and μ measurement. A good reference colloid should exhibit (a) excellent stability over the course of the ME experiment, (b) a surface potential as insensitive to pH as possible, (c) commercial availability in a highly pure and reproducible form, and (d) resistance to environmental factors that could change its properties upon storage. A reference colloid satisfying these conditions would provide for reproducible μ measurement and thus would be useful for standardizing the ME instrument. Polystyrene (PS) latex fulfills most of the criteria cited above (except possibly (d)). Furthermore, it has been used widely for electrokinetic studies for decades.[6,7] Features of PS latex make it suitable for use as a reference colloid and procedures for obtaining accurate and precise μ measurements for the reference colloid and silicon nitride powder will be discussed.

EXPERIMENTAL PROCEDURE

All chemicals* other than those cited below were ACS reagent grade or better, and were used as-received. Deionized water with a resistivity of 18 MΩ•cm was obtained with a Milli-Q Water System.[†] Sulfate group-stabilized 0.5 μm PS latex was purchased from three sources.[‡] Suspensions of PS latex were prepared by diluting a concentrated commercial dispersion into a desired volume fraction with electrolyte solution.

Surface properties of silicon nitride powder (UBE SN E-10)[§] were studied by ME as a function of processing history. Powders were examined as-received (R-SN) or freeze-dried[¶] in various ways. Suspensions of 10 vol% R-SN powder sonicated for 15 min, subsequently stirred 15 min, and finally aged 30 min were freeze-dried in water (FD-SN). Powders doped with aluminum or yttrium were prepared by freeze drying R-SN aqueous suspension containing nitrate salts of aluminum or yttrium. Concentrations employed for both aluminum and yttrium dopants were 1 wt% hence referred to as Al-FD-SN and Y-FD-SN, respectively.

---

*   Aldrich Chemical Co., Inc., Milwaukee, WI.
†   Millipore Corp., Bedford, MA.
‡   Polysciences, Inc., Warrington, PA, Interfacial Dynamics Corp., Portland, OR, and Sigma Chemical Co., St. Louis, MO.
§   UBE Industries, Inc., New York, NY.
¶   Dura-Dry, FTS Systems, Stone Ridge, NY.

Suspensions for ME were prepared by ultrasonically dispersing the 0.01 vol% powder in 0.01 M KNO_3 solution. Suspension pH was adjusted by either HNO_3 or KOH solution, after which suspensions were aged in sealable Falcon polypropylene containers.[*] Prior to characterization, each suspension was divided into two groups for μ[†] and pH[‡] measurements. An alternating electric field with the strength of 15 V/cm was used for ME. In order to acquire reproducible measurements, the ME cell was preconditioned for about 3 min using sample suspension prior to the actual measurement. The μ and pH measurements were performed simultaneously at 25°C for each sample. At least seven measurements were made and subsequently averaged together to obtain each μ data point, except for time effect studies, which utilized a single measurement per data point.

RESULTS AND DISCUSSION

Of the three latices described above, the Polysciences latex was selected as a reference because it displayed a high zeta potential (ζ~100 mV) and its mobility was the least sensitive to changes in pH for a given electrolyte concentration. For instance, when a 0.01 M KNO_3 electrolyte was employed over a pH range of 5.5 to 10.5, μ fell within the range of $(-7.71\pm0.05)\times10^{-8}$ m^2/Vs.

The pH insensitivity made the Polysciences latex colloid optimum as a reference material. As a result, additional experiments were necessary to verify the operating condition of the ME equipment. The condition of the cell and the associated optics can be examined by using the reference colloid to check the stationary layer positions of the cell.[8] This was accomplished by varying solution pH to vary the electro-osmotic velocity[8] of the suspending solution. Using the intersection of two or more cell profiles yielded experimental stationary layer positions within 5 μm of the position determined from fluid mechanics. The symmetrical μ profiles also showed that there were no dimensional asymmetries present. The steepness of the μ profile increased with pH; thus, when the stationary layer position was not properly located, μ was more subject to instrumental error.

Assuming the instrument is operating properly, obtaining reliable μ values for the reference colloid also depends on the sample preparation conditions. First, our studies revealed that a solids loading greater than 0.005 vol% was required for both PS and silicon nitride powders before a constant μ could be obtained (Fig. 1). Using different instrumentation,[§] this has also been found for calcite powders.[9] A decrease in the absolute value of μ has been attributed to hydrodynamic interactions[10] and colloidal phenomena such as double layer compression. In this experiment, how-

---

[*] American Scientific Products, Edison, NJ.
[†] Zetasizer II/4700c correlator, Malvern Instruments, Malvern, U.K.
[‡] Brinkmann 686 Titroprocessor; Metrohm Ag/AgCl double junction electrode; Brinkmann Instruments Co., Westbury, NY.
[§] Laser Zee, Model 500, Pen Kem, Inc., Bedford Hills, NY.

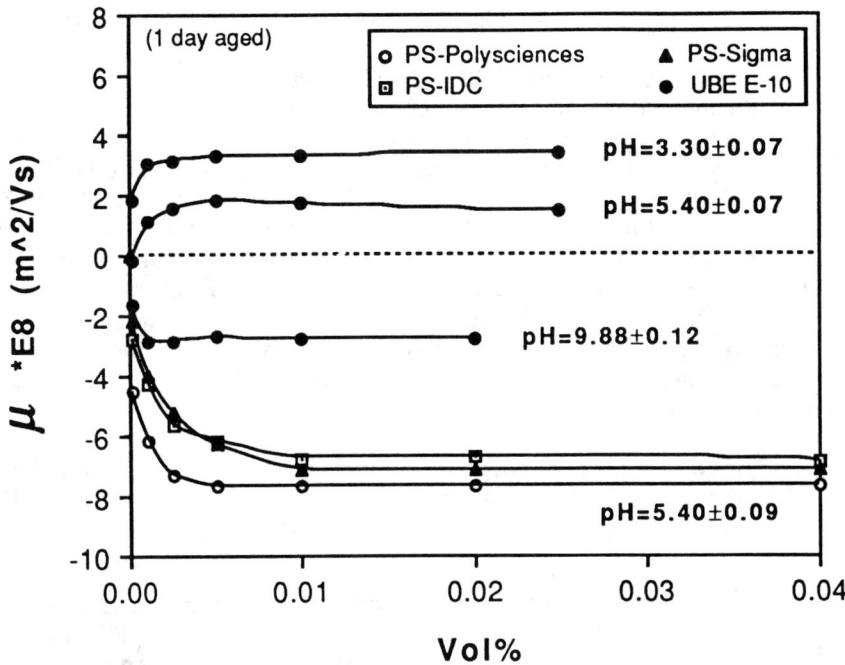

Figure 1 Electrophoretic mobility as a function of solids loading in 0.01 M KNO$_3$ solution.

ever, the absolute value of μ was found to increase. This difference could have been caused by insufficient particle population in the sampling volume for laser Doppler velocimetry. Second, the mobility was found to vary with the time elapsed[8] after introducing the sample into the ME cell regardless of the solution pH, especially in the first two minutes. For example, in 0.01 M KNO$_3$ solution, the latex μ increased from -7.12 to -7.46x10^{-8} m^2/Vs for pH at 4.31 and from -7.42 to -7.73x10^{-8} m^2/Vs for pH at 9.67. The magnitude of the μ change also varied with the concentration of electrolyte.[8] The higher the electrolyte concentration used, the larger magnitude in μ change was observed. The time-dependent behavior of the latex μ could be due to the dynamic equilibrium of the distorted electric double layer under an applied alternating field.Third, the μ was also found to vary with the particle size. The average of the μ differences for the 0.5 and 1.0 μm latices (i.e., $|\mu_{0.5}-\mu_{1.0}|$) was (0.44$\pm$0.08)x10^{-8} m^2/Vs which were calculated for the entire μ vs. pH curves. The typical μ value for 1.0 μm latex in pH-insensitive range (i.e., pH>5.5) was -8.04x10^{-8} m^2/Vs and was -7.68x10^{-8} m^2/Vs for 0.5 μm one, respectively. The 1.0 μm latex consistently has a higher surface charge than the 0.5 μm one. Therefore, all the measurements were performed immediately after injection into the conditioned sample cell (~2 s). This overall approach led to the collection of reproducible μ data.

After using the above reference colloid to check our instrument, μ studies on R-SN powder were conducted. In general, neither time dependency nor electrolyte effects were observed for this powder. To examine μ as a function of aging time, 0.01 vol% solids loading suspensions were stored in a sealed container for a period of 1 to 30 days. The IEP of 0.01 vol% UBE powder in 0.01 M KNO_3 solution shifted from 7.4 to 8.8 as it was aged for 30 days (Fig. 2). A discontinuity over the 6 to 8 pH range was observed for samples aged for more than 1 day but less than a month. In general, longer aging times shifted the IEP to higher values. This phenomenon could be due to leaching of soluble silica (e.g., silanol groups) from silicon nitride surfaces during aging.[4] At high pH, the leaching process leaves behind more basic silicon groups (e.g., $-SiNH_2$ or $-Si_2NH$) on powder surfaces than acidic silanols yielding a more basic surface.

Figure 3 shows that the FD-SN powder exhibited a lower IEP than the R-SN one after 1 day of aging in aqueous solution (Fig. 3). This can be explained by the supersaturation of the soluble silica during the freeze-drying of concentrated R-SN

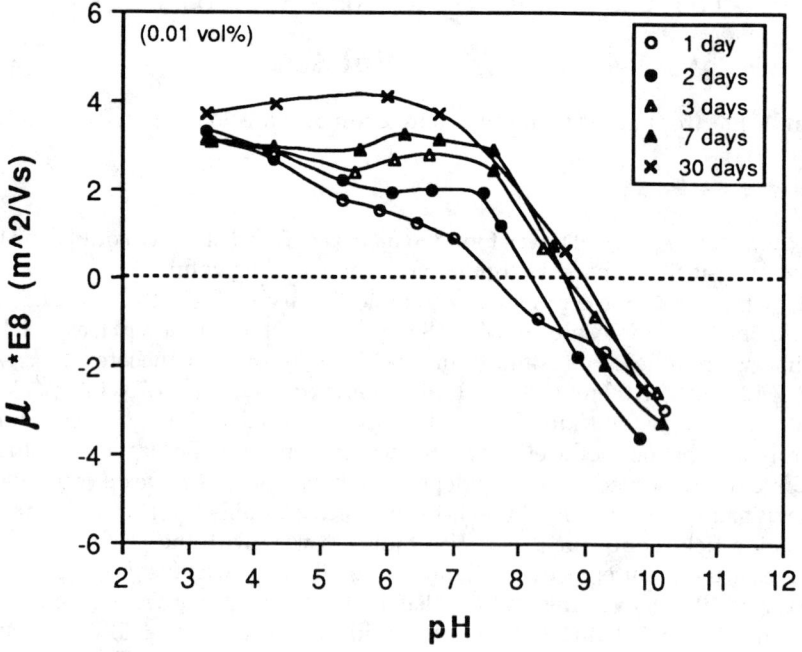

Figure 2 Electrophoretic mobility of 0.01 vol% silicon nitride powder in 0.01 M KNO_3 as a function of aging time.

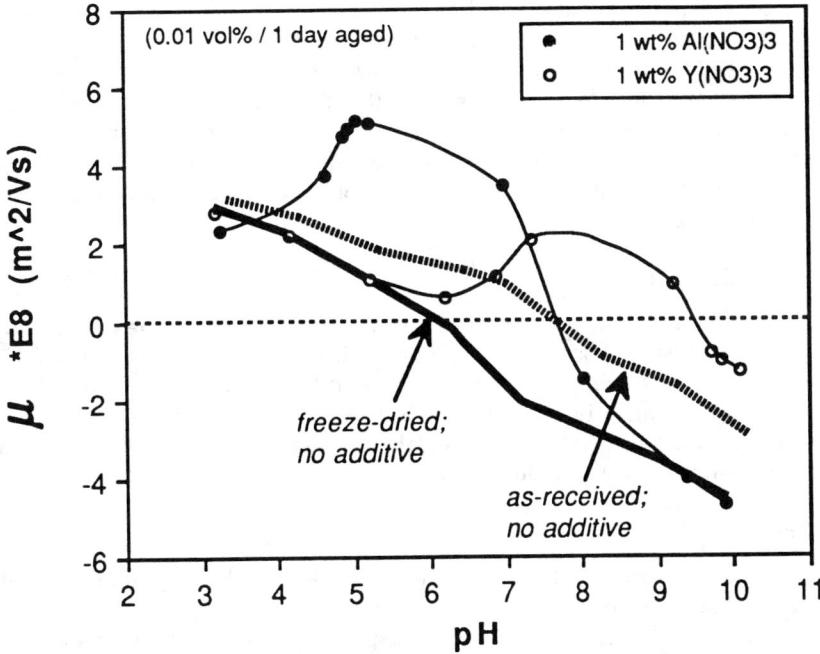

Figure 3 Electrophoretic mobility of 0.01 vol% silicon nitride powder in 0.01 M KNO₃ as a function of dopant and preparation history.

slurry, which causes the reprecipitation of the leached silica back on the powder surfaces. The reprecipitated silica changes the surface composition to have relatively more silanol than amino groups. A significant change was observed for μ-pH curves starting from pH 4.5 for Al-FD-SN and from pH 7 for Y-FD-SN (Fig. 3). The Al- and Y-FD-SN powders showed the μ-pH curves typical of adsorption of hydrolyzable metal ions at the oxide-water interface.[11] At higher pH values, the μ-pH curve followed the curve of the corresponding hydroxides precipitating from the solutions on silicon nitride powder surfaces.

CONCLUSIONS

Based on criteria cited earlier, polystyrene latex has been identified as a good reference colloid for establishing the proper operation of a microelectrophoresis apparatus. However, the effects of solids loading, time-dependence, particle size, and electrolyte concentration on μ must be determined before reference μ values can be established for the colloid. Neither time dependency nor electrolyte effects were observed for silicon nitride powder. However, the solids loading effect was

245

found on both polystyrene latices and silicon nitride powder for solids loadings less than about 0.005 vol%. The IEP of silicon nitride powder increased as the aging time increased due to leaching of soluble silica from silicon nitride surfaces. The freeze-dried silicon nitride powder has lower IEP (6.0) than as-received powder (7.4) after 1 day of aging caused by the reprecipitation of the silanol groups back on the powder surfaces. The surface property of the doped silicon nitride powders changed significantly starting from pH 4.5 using aluminum nitrate and from pH 7 using yttrium nitrate.

ACKNOWLEDGMENTS

The authors would like to acknowledge the generous support of the U. S. Department of Energy, Assistant Secretary for Conservation and Renewable Energy, Office of Transportation Systems, as part of the Ceramic Technology for Advanced Heat Engines Project of the Advanced Materials Development Program, and also the Center for Ceramic Research at Rutgers University, and the New Jersey State Commission on Science and Technology.

REFERENCES

1. J. S. Reed, Introduction to the Principles of Ceramic Processing; p.141. John Wiley & Sons, Inc., New York, 1988.
2. G. A. Parks, "Aqueous Surface Chemistry of Oxides and Complex Oxide Minerals," Adv. Chem. Ser., 67, 121-160 (1967).
3. G. A. Parks, "The Isoelectric Points of Solid Oxides, Solid Hydroxides, and Aqueous Hydroxo Complex Systems," Chem. Rev., 65, 177-198 (1965).
4. L. Bergstrom and R. J. Pugh, "Interfacial Characterization of Silicon Nitride Powders," J. Am. Ceram. Soc., 72 [1] 103-109 (1989).
5. P. K. Whitman and D. L. Feke, "Comparison of the Surface Charge Behavior of Commercial Silicon Nitride and Silicon Carbide Powders," J. Am. Ceram. Soc., 71 [12] 1086-1093 (1988).
6. R. H. Ottewill and J. N. Shaw, "Electrophoretic Studies on Polystyrene Lattices," J. Electroanal. Chem., 37, 133-142 (1972).
7. B. J. Marlow, D. Fairhurst, and W. Schutt, "Electrophoretic Fingerprinting and the Biological Activity of Colloidal Indicators," Langmuir, 4, 776-780 (1988).
8. J.-F. Wang, R. E. Riman, and D. J. Shanefield, "Reliable Electrokinetic Characterization Procedures for Ceramic Powders"; pp.293-298 in Proceedings of the Better Ceramics Through Chemistry IV of Materials Research Society, San Francisco, CA, 1990.
9. B. Siffert and P. Fimbel, "Parameters Affecting the Sign and the Magnitude of the Electrokinetic Potential of Calcite," Colloids and Surfaces, 11, 377-389 (1984).
10. J. L. Anderson, "Concentration Dependence of Electrophoretic Mobility," J. Colloid Interface Sci., 82 [1] 248-250 (1981).
11. R. O. James and T. W. Healy, "Adsorption of Hydrolyzable Metal Ions at the Oxide-Water Interface," J. Colloid Interface Sci., 40 [1] 53-64 (1972).

INDEX

Acoustophoretic mobility, 31, 38
Adams, R.W., 157
Agglomerate
  binding strength, 66
  size, 66
  strength, 125
Agglomeration, 1, 54
Aggregation, 54
Alcohol, 165
Alkoxide synthesis, 17
Alumina, 24, 31, 141, 178, 197, 205
  alpha alumina slips, 81
  coating, 31
  granular films, 88
  homogeneous distribution, 31
  powder, 17
  zirconia-toughened composites, 24
Aluminum monohydroxide, 54
Ansari, R.R., 54
Aqueous media, 24

Barclay, D.A., 81
Barium
  oxalates, 8
  titanate, 8, 115
  titanyl oxalate, 8
Beryllia
  ceramics, 211, 231
  substrates, 211
Binary solvent, 197
Binder, 165, 191, 205
  removal, 115
Bingham behavior, 81
Borosilicate glass, 197
Burnfield, K.B., 191

Caley, W.F., 46
Capillary forces, 95
Carlson, W.B., 132
Cawley, J.D., 54
Cellulose ethers, 191
Ceramic, 101, 165
  complex, near-net-shaped, 189
  glass, 197

high performance, 157
  multilayer, 115
Cherng, C-L., 197
Chiu, R.C., 88
Cima, M.J., 8, 88, 115
Cold isostatic pressing (CIP), 231
Colloidal
  powder treatments, 189
  processing, 125
  stability, 1
Complex shape, 157
Connell, S., 17
Controlled stress rheometer, 81
Copper oxalates, 8
Cracking, 108

Darcy's law, 187
Davis, W.M., 8
Deformation, 115
Delamination, 115
Dilatant fluid, 81
Dispersion, 197
  structure, 81
Donut-shaped granules, 46
Drop formation, 46
Drying, 88, 101, 108

Edirisinghe, M.J., 165
Electrokinetic behaviour, 24
Electron beam perforation, 219
Enhanced reliability, 189
Esposito, L., 178
Extrusion, 132

Films
  cracking, 88
  granular alumina, 88
Filtration kinetics, 187
Finite difference method, 132
Flaws, process related, 125
Flow curves, 81
Foam, 149
Forsterite, 197
Fractal, 54

Fracture mechanics, 108

Gelcast, 101
Goski, D., 24
Green
  sheet, 205
  tape, 197
Grooves, 225

Haber, R.A., 73, 172
Hallock, R.B., 8
Hardy, A.B., 141
Hesse, K., 17
Hydroxyl content, 197

Inclusion morphology, 95
Injection molding, 157
Isopressing, 231

Kendall, K., 125
Key, K.C., 95
King, H.W., 46
Kinsman, K., 17
Konsztowicz, K.J., 24, 46
Kottman, R.E., 211
Kovell, G.A., 231
Kwak, J.C.T., 24

Lange, F.F., 187, 189
Latex, 240
Li, X., 108
Lin, J-C., 197
Linear viscoelastic region, 81
Liquid stream necking, 46
Look, J-L., 1
Low cost fabrication, 157
Low pressure, 157
Luminescent material, 17

Machining, 231
  of green ceramic material, 225
Maksym, G., 46
Maksym, T., 46
Malghan, S.G., 31, 38
McKittrick, J., 17
Meenan, B.J., 172
Meyer, W.V., 54
Microelectrophoresis, 240

Minnear, W.P., 149
Mishiro, S., 225
Molding, 165
  injection, 157
Morphology, 1
Moudgil, B.M., 66
Multilayer ceramic, 115
Murakami, H., 219

Nagata, K., 205
Nakabayashi, H., 225
Nanoindenter, 125
Net shape dimensions, 157
Niesz, D.E., 172
Novich, B.E., 157

Omatete, O.O., 101
Oscillatory measurements, 81
Oxalates, copper, barium, yttrium, 8

Packing
  density, 187, 205
  particle, 115
Pair potentials, 1
Patching, M., 165
Pei, P.T., 38
Percolation
  limit, 95
  structures, 197
Perforation, high-speed, precise, 219
Peterson, B.C., 191
Plastic
  bodies, 132
  processing, 125
Polarity, 197
Polyacrylic acid, 31
Polymer
  adsorption, 205
  superabsorbent, 141
Polymethacrylic acid, 81
Polystyrene, 240
Polyvinyl
  alcohol, 125
  butyral, 197
Power law viscosity, 132
Precipitation, 1, 8
Prereaction, 172
Pressure

filtration, 187
slip casting, 178
Processing, 191
colloidal, 125
plastic, 125
Pseudoplasticity, 81, 197

Reed, J.S., 132
Reference colloid, 240
Relic process, 141
Rheological
behavior, 73, 205
measurements, 81
Rheology, 197
cake, 73
powder compact, 189
Rhine, W.E., 8, 141
Rib holes, 225
Rim, Y.H., 54
Riman, R.E., 240

Sakurai, K., 219
Salomoni, A., 178
Sasaki, S., 219
Self-reinforced, 172
Sepulveda, J.L., 211, 231
Shanefield, D.J., 240
Shear
hardening, 81
thickening, 81, 197
thinning, 81, 189, 197
Short-range repulsive hydration layers,
187
Silica, 54, 141
Silicon
carbide powder, 38
nitride, 31, 172, 240
nitride powder, 38
Sintering aids, 31
Sivakumar, A., 31
Slip casting, 95, 178
Slurry, 197
coagulated, 187
concentrated, 66
dispersion of, 66
surface charge of, 66
viscosity of, 66
Sluzky, E., 17

Solid oxide fuel cell, 125
Solvent polarity, 197
Spray dried agglomerates, 125
Spray drying, 46
Springback, 46
Springgate, M.E., 66
Stamenkovic, I., 178
Storage modulus, 81
Strehlow, R.A., 101
Stress, 88
Substrate, 197
Sundback, C.A., 157
Superconductors, high T_c, 8
Surface
charge modification, 31
oxide thickness, 38
tension, 46
Suspension, 205
Suzuki, K., 225

Takeno, S., 219
Tang, Y., 115
Tape
casting, 191, 197, 205, 211
substrates, 211
Tensile testing, 191
Titanium alkoxides, 1
Tomkins, K.L., 165
Tsao, I., 73
Tucci, A., 178

Uematsu, T., 225
Ultrasonic vibration tool, 225
Uniform particles, 1

Vargha-Butler, E., 46
Velamakanni, B.V., 187, 189
Viscoelastic systems, 81
Viscosity, 46, 197
dynamic, 81
power law, 132

Walls, C.A., 101
Wang, C-M., 197
Wang, J-F., 240
Wang, P.S., 31, 38
Weakly attractive particle network, 189
Weiser, M.W., 95

X-ray
crystallography, 8
photoelectron spectroscopy, 31, 38

YAG powder, 17
Yamane, Y., 219
Yeh, T-S., 197
Yield point, 81
Yttria-stabilized zirconia, 24, 125
Yttrium oxalates, 8

Zeta-potential, 24
Zheng, J., 132
Zirconia, 24, 141, 178
membrane, 125
powder, spray dried, 125
strength, 125
-toughened alumina composites, 24
Zukoski, C.F., 1